Universitext

Jacques Istas

Mathematical Modeling for the Life Sciences

With 31 Figures

Jacques Istas
Départment IMSS BSHM
Université Pierre Mendès-France
38000 Grenoble
France
e-mail: jacques.istas@umpf-grenoble.fr

Based on the French edition "Introduction aux Modélisations Mathématiques pour les Sciences du Vivant", Mathématiques et Applications, Vol. 34, Springer-Verlag 2000

Mathematics Subject Classification (2000): 92B05

Library of Congress Control Number: 2005926252

ISBN-10 3-540-25305-X   Springer Berlin Heidelberg New York
ISBN-13 978-3-540-25305-1   Springer Berlin Heidelberg New York

Springer is a part of Springer Science+Business Media
springeronline.com
© Springer-Verlag Berlin Heidelberg 2005
Printed in Germany

Cover design: Erich Kirchner, Heidelberg
Typesetting by the author using a Springer LaTeX macro package

Printed on acid-free paper   41/sz - 5 4 3 2 1 0

Contents

Notations

- $\mathbb{R}$: set of real numbers.
- $\mathbb{N}$: set of nonnegative integers.
- $\mathbb{Z}$: set of integers.
- $C_n^k = \dfrac{n!}{k!(n-k)!}$: binomial coefficient.
- $\#A$: cardinal number of A.
- $\mathbf{1}_A$: indicator function of A.
- $L^2(\Omega)$: set of square integrable function on Ω.
- $\dfrac{\partial f(x,y)}{\partial x}$: partial derivative of function f with respect to x.
- $gradf \equiv \dfrac{\partial f}{\partial x}\mathbf{i} + \dfrac{\partial f}{\partial y}\mathbf{j} + \dfrac{\partial f}{\partial z}\mathbf{k}$ (gradient of f).
- $\nabla J \equiv \dfrac{\partial J}{\partial x} + \dfrac{\partial J}{\partial y} + \dfrac{\partial J}{\partial z}$ (divergence of J).
- $\Delta f \equiv \dfrac{\partial^2 f}{\partial x^2} + \dfrac{\partial^2 f}{\partial y^2} + \dfrac{\partial^2 f}{\partial z^2}$ (Laplacian of f).
- $\mathbf{E}X$: expectation of the random variable X.
- $\mathbf{P}(\omega)$: probability of the event ω.
- $\operatorname{var}(X)$: variance of the random variable X.
- i.i.d. r.v.: independent and identically distributed random variables.
- g.c.d.: greatest common divisor.

1

General introduction

1.1 Preface

Proposing a wide range of mathematical models that are currently used in life sciences may be regarded as a challenge, and that is precisely the challenge that we are going to take up. Of course, this panoramic study does not claim to offer a detailed and exhaustive view of the many interactions between mathematical models and life sciences. Some topics will not be dealt with in this book. If they are not, it is usually because the amount of mathematical tools has no common point with the biological application; or because the topic is marginal, not very convincing or obsolete; or sometimes because the required mathematics are not available yet. However, enough of what we are not going to deal with! Let's get on with what we are actually going to look at.

We are proposing an introduction to mathematical models in life sciences. Before doing maths, we first need to model, which is not an easy business. What is a (good) model? The first purpose of a model is to highlight some important and general phenomena. As a consequence the model must be simple, sometimes a caricature. *Pluralitas non est ponenda sine necessitate*[1], it will then look like a paradigm. The model must enable one to foresee the behavior of the system it is supposed to represent, from a quantitative viewpoint if possible, from a qualitative viewpoint if not. However, simplicity is what the person in charge of modeling should constantly aim at, even if we run the risk of losing (partially but inevitably) realism. That is the viewpoint adopted here. We will not insist on a precise and detailed representation of reality. We will rather offer in parallel some models (which will help understand a phenomenon), and real examples (which will explain). Nevertheless, we must not forget that this book is mathematically oriented. The discussion on models from a biological or ecological viewpoint is drafted rather than

---
[1] Plurality should not be assumed without necessity, William of Ockam 1285?–1349?

actually detailed. On the other hand, we will try to initiate the non-biologist into biological problems and tools.

What do we want to model? "Ecology" and "biology" are very restrictive terms; "life sciences" is wider but certainly lacks precision. Some topics are necessary, either because of their historical interest when they are seminal models, or because of their practical use. That is why many of the models studied here come from demography and ecology. However, other prospects have been considered: the theory of evolution, quantitative genetics, DNA sequencing ... but it is obviously a subjective choice and we abide by this rule.

Having defined our field, which is life sciences, and our aim, which is modeling, we will now turn to the purely mathematical part of this book, albeit with discernment. Indeed, the underlying biological context will certainly not be pushed into the background in favor of mathematics. Issues that, for the mathematician, are of prime importance (for instance, the issue of the existence or uniqueness of the solutions of a partial differential equation), will often be considered as secondary by the biologist. A difficult balance between mathematical rigor and biological interest must therefore be found. The pursuit of this balance may sometimes seem disconcerting to the reader; for example theorems, which cost mathematicians significant efforts, will sometimes be presented rapidly and without any proof. However, we will detail calculations that, from a mathematical point of view, are much more down-to-earth but that will help understand the underlying biological issue.

1.2 Structure of the book

Our main table of contents is quite classical and is based on the dichotomy between determinist and stochastic models. The first part deals with determinist models only. The second part deals with stochastic models and also with the comparisons, when possible and legitimate, between those two approaches. Our aim will not be ideological or philosophical. We are not trying to "sell" either the determinist or the stochastic model. However, we will try to find which of the two is the most appropriate.

Problems of population management began to pave the way for population dynamics as an object of study, especially its confrontation of paradoxical situations that reason could not resolve. For example, why is it the decline of fishing in the Adriatic during World War I later resulted in poorer catches and in a relative abundance of predators? Similarly, why should the successful use of a noxious insecticide on some forest insects have been followed afterwards by a unprecedented rapid multiplication of those insects?

Foreseeing the effect of an action on an ecological system is not a trivial exercise. Neither is acting on it intelligently. Such issues take on a powerful significance nowadays because mankind has many powerful means of action at its disposal and uses them without restriction in order to alter nature. How are we to save the biological treasures of our planet then? How will we manage to extract our biological resources in an efficient and long-term way? We will try, when possible, to introduce the reader to these topical issues through discussions on the models we have chosen as examples.

Population dynamics is a study of complex dynamical systems. Its complexity is due to the richness of relations that link any population to its environment and to other populations. It is also due to the fact that we do not, and will not, master the details of these interactions. The main problem is not to be distracted from our aim by too many details but, on the contrary, to highlight what is necessary to understand the studied phenomena. Population dynamics rapidly leads to models that, while simple to formulate, may be difficult to study and may display a multitude of behaviors. The first two chapters will deal with the main themes: historical yet still topical issues (the temporal evolution of one or several populations, interactions and co-existence between populations, space-time distribution of these populations, age distribution, and so on ...) or more recent issues, such as the chaotic dynamics of insect populations, which are problematic when considering the management of biological resources as we know it.

Population dynamics is as ancient as the study of the conflicts within a species is recent - it is merely a matter of decades. Its sources of inspiration generally lie in the theory of evolution, in which we deal with the co-existence of antinomic behaviors within a population. How can one model the selective value of a behavior? Can we use the term "model" when dealing with ethology?

The problem of the hawk and the dove is an outline of mathematical theory on the behavior of animals, in which we try to understand how aggressive and peaceful behaviors are both viable.

Curiously enough, the study of gender distribution within a species or, to repeat the famous question "Why sex?", is similar to the mathematical apparatus of the model of conflicts. It is the game theory, though initially stated to model economic facts. Here lies a good example of the universal power of mathematics. The short chapter devoted to game theory will close the study of purely determinist models.

Let us now turn our attention to the stochastic problem. Considering the amazing complexity of living beings, some may wonder why every such model is not stochastic. Others will deny this, only regarding the stochastic model as a fake generalization of a determinist problem. That is why we will begin our chapter on Markovian models with a simple but convincing example. Let us imagine a desert island where a group of shipwrecked people (dark-haired and fair-haired) have landed. Several generations follow and a small colony eventually populates the island. What is the hair color of the initial shipwrecked people's descendants? The determinist will answer as would Hardy and Weinberg: the number of dark-haired and fair-haired people would remain unchanged. The stochastician would see that his Markov chain has been absorbed and that the inhabitants would have the same hair color. Reality would prove this to be correct. However, the Markovian model does not state what color it is. Every model has its limits ...

Let us now travel from our desert island to Sahelian Africa. Cereals have been cultivated there for a long time. How can one account for the fact that man could one day domesticate cereals from wild species in an environment that is far from adapted to agriculture? Such an example is not only interesting from a historical point of view, it is also valid for the problem of genetically modified organisms. How can the resettlement of genes between neighboring populations be modeled? The domestication of pearl millet in Sahelian Africa gives us a simple mono-allelic model where dynamical systems and Markov chains co-exist. This will be studied thoroughly.

What is the probability of a family name disappearing? Such a simple question, similar to those concerning genealogy, gave birth to branching processes. A branching processes is a random family tree. Its study is not very difficult on the whole and it will be studied thoroughly. Genealogy is a worthy subject but its practical interest is not sufficient, in our opinion. We will apply the branching processes to a modern technique of DNA duplication - the question is to determine how many DNA strands were originally present. Branching processes are sometimes too restrictive to model temporal phenomena. A short study of percolation will therefore be undertaken before studying the spatial branching processes. Indeed, the main interest of the branching processes seems to lie elsewhere, in an extension of such processes, those of spatial branching. These are temporal branchings that are linked to an additional law of people dispersion.

After the first glaciation, oak trees colonized throughout Europe at high speed. However, anyone observing such trees will notice that their acorns fall at their very roots, except those that are taken away by birds. How can we account for the swift progress of the oaks throughout Europe? The spatio-temporal determinist model, based on reaction-diffusion equations, does not enable us to account for this phenomenon in a satisfying way, whereas the spatial branching processes, thanks to a subtle mix of large deviations and exponential growth, provide a trustworthy model for the oak colonizing process. Who could ever predict that the survival of the species was made possible thanks to jays?

Last but not least, let us turn to statistics. Indeed, when speaking of stochastic models, one can also think of parameter estimates, confidence intervals and hypothesis tests. It would have been possible to deal with statistics in an appendix but that would amount to considering statistics as an ancillary subject, only of value to confirm results we already foresaw. We have intentionally devoted a proper chapter to statistics. One could object that statistical methods are not models, strictly speaking. This sounds acceptable but statistical methods are often necessary to account for numerous biological issues. Hence, rather than sticking to basic but rather dull statistical problems, we have opted for examples whose core is statistical. In this way, the likelihood method will be illustrated with the search for a gene on the DNA sequence, and the likelihood theory will be applied to weevil life.

We have added three appendices: one on ordinary differential equations, another on evolution equations and the last one on probability to make this book an autonomous work.

A general bibliography is available at the end of the book and at the end of each chapter a more specific bibliography can be found.

1.3 Acknowledgments

This book comes originally from a French book entitled "*Introduction aux modélisations mathématiques pour les sciences du vivant*" (Mathématiques & Applications, vol. 34, (2000), Springer-Verlag).

Many colleagues, but also friends helped us with an example, a figure, a reference, a review of the draft or a translation. Our thanks go to Valérie Bénéfice, Etienne Bertin, Eric and Susana Bonnetier, Sally Brown, Frédérique Clément, Denis Couvet, Gersende Fort, Thierry Gallais, Pierre-Henri Gouyon, Marie-Anne Poursat, Christine Jacob, Etienne Klein, Sophie Lambert, Catherine Laredo, Alain Latour, Frédérique Letué, Jérome Renault, François Rodolphe, Sophie Schbath, Philippe Souplet and Nicolas Trotignon. Special thanks to Nathalie Guillot-Perdoux for her translation support.

2

Continuous-time dynamical systems

2.1 Introduction

The demography of animal species has always interested biologists. It was especially seen as a first approach towards human demography. Demography necessarily appeals to mathematics, even on a very basic level. In fact, it represents a natural crossroads between maths and biology. From a historical point of view, the first models applied to demography were based on dynamical systems. It seems logical, then, to start this book with a survey of elementary demographical models since they constitute the first step towards more elaborate systems.

The first question to be answered concerns the evolution of a population in time. The naive model in which reproduction and death rate are proportional to the number of people leads to an explosion or an extinction of this population at an exponential rate, according to the parameters. This approach, stated by Malthus (though earlier suggested by Euler), is unrealistic globally speaking, at least in the long term. We will thus have to introduce a corrective term that allows the members of an isolated population to converge towards a constant number. The idea is to prove that there exists an ideal number such that if the environment is not altered, the size of the population will stabilize around this ideal number. This is what the logistic model proposes. It is a very simple model that only depends on two parameters, which have a quite clear biological interpretation. We will find this logistic model in many of the models throughout the book.

When only taking the temporal evolution of a population into account, we are led to drastic simplification. Such is the case when we suppose that people are sexually mature as soon as they are born. It is not difficult to build a model that accounts for the gap between birth and sexual maturity. It leads to partial differential equations. Studying the solutions of these equations and their qualitative behavior can be quite complicated and is out of the scope of this book: we will only give some basic results of this theory. This is the first illustration of what was said in the main introduction, that is to say, the model has to be simple, even caricatural, and realistic models are not necessarily the best.

Is an epidemic due to a major alteration of the environment? How can we alter the environment so as to fight against an epidemic? We will take a well-known example, that of the spruce budworm. This model will show that a very small alteration in the environment (here, the number of predatory birds) can cause an epidemic. It will also show that once it has spread, coming back to a normal situation is very difficult.

Formally speaking, the model of the spruce budworm only involves one population. However, two other populations actually interact with the spruce budworms: the parasitized spruces and predatory birds. This model assumes the simplest interactions between spruces, caterpillars and birds. However, we clearly feel that the interactions between various populations must be modeled with more precision. The Lotka-Volterra model is the historical example of a model between a population of preys and a population of predators. While in a one population model, the size of the population converges to a constant value, two-population models may show a different behavior: periodical cycles. Such cycles can be observed in nature and thus justify the model applied to the predator-prey systems. Another commonplace interaction between populations is surveyed next because of its significance in the definition of the concept of ecological niche. It is the competition model, in which two populations share the same resource but with conflicting episodes.

So far we have neglected the spacewise aspect of the chosen populations. To this end, we will first write conservation equations for populations living both in space and time. Such equations are special partial differential equations, called reaction-diffusion equations. See Appendix A.2 for an introduction to the existence, uniqueness and boundedness of the solutions to reaction-diffusion equations. We will devote this section mainly to a particular mathematical aspect: traveling waves, *i.e.* particular solutions that evolve in time without changing their shape. We will study the traveling waves of the Fisher equation (in fact the spatial logistic model) from a qualitative point of view. Traveling waves, among other things, enable one to model the geographical spreading of an epidemic and to understand how an epidemic can spread, without any impulse, toward a precise direction. Traveling waves can be found in nature or, rather, there are propagative phenomena in nature, which can be modeled with traveling waves, even if their shapes is only roughly preserved in time. There again, our purpose is not to legitimize the solution to the reaction-diffusion equation with a maximum of veracity, but to emphasize an existing qualitative phenomenon. The spreading of the larch bud moth along the Alpine arc will give us an interesting illustration of traveling waves.

2.2 Historical demographical models

2.2.1 Basic models

The most basic models deal with one single species. They provide with a convenient starting point for more general models.

Firstly, we will study the growth of a population in terms of *time* and second, its *spatial repartition*.

Let $N(t)$ be the number of individuals in the species at time t. We assume that $N(t)$ is large enough; $N(t)$ is a real number and not an integer. This assumption is not problematic when $N(t)$ is effectively huge. On the other hand, what does $N(t)$ small mean? Does it mean that $N(t)$ is (mathematically) close to zero, or that $N(t)$ is equal to some unit? We will see that some wise change of parameters usually allows to give a precise meaning to "$N(t)$ small".

The population dynamics is described by a *conservation equation*:

$$\frac{dN(t)}{dt} = births - deaths + migrations .$$

Malthusian model

In the most basic model, there are no migrations, and births and deaths are proportional to the population $N(t)$.

$$\frac{dN(t)}{dt} = \phi N(t) - \mu N(t)$$
$$= rN(t) ,$$

where ϕ and μ are positive constants. We can easily deduce that $N(t) = N(0) \exp(rt)$. So, if $r > 0$, the population grows exponentially. If $r < 0$, the population decays exponentially: from a biological point of view, the species disappears.

There is a borderline case when $r = 0$: population size remains constant. This solution exists from a mathematical point of view but is unrealistic from a biological point of view. This is why we avoid the study of such borderline cases in this book.

Logistic model

It seems reasonable to include the effects of the environmental resistance when the population grows in the previous model. Basically, there exists an "ideal" population size, called the *carrying capacity*. Below the carrying capacity, the population grows. Above it, it decreases. [83] proposed the *logistic model*:

$$\frac{dN(t)}{dt} = rN(t)\left(1 - \frac{N(t)}{K}\right), \tag{2.1}$$

where r and K (the carrying capacity) are two positive constants.

The logistic model is rather basic. However, in some cases its predictions might be accurate: the Belgian Pierre Verhulst, 1804-1849, predicted that the Belgian population would stabilize around 9,4 millions of inhabitants, close to its current size (10,1 millions in 1994). However, Verhulst did not take into account immigration, the death toll of the wars, the drop of the birth rate ... It might be a matter of chance that its prediction seems so good! Thus one should avoid coarse models like the logistic model in order to make quantitative predictions.

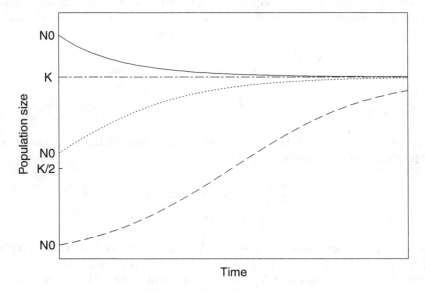

Fig. 2.1. Logistic curves

The equation (2.1) has two steady points $N = 0$ and $N = K$ where $\frac{dN(t)}{dt} = 0$ [1]. A linearization of (2.1) near $N = 0$ proves that the steady state $N = 0$ is unstable. The linearization near $N = K$ proves that the steady state $N = K$ is stable.

A little algebra leads to the analytic solution of (2.1):

$$N(t) = \frac{N(0)K\exp(rt)}{K + N(0)(\exp(rt) - 1)},$$

and its graph is given in figure 2.1. We can check that the population converges to K as $t \to +\infty$: K is indeed a carrying capacity.

---

[1] See appendix A.1 for classical results on ordinary differential equations.

2.2.2 General case

A general demographical model for one single species is an autonomous differential equation in the form $\dfrac{dN(t)}{dt} = f(N(t))$, where f is a function of N.

The steady states are given by the solutions of the equation $f(N) = 0$. Usually, 0 is a steady state since there is no spontaneous generation. Let $N^\star$ be a steady state. Let us find the behavior of the solution $N(t)$ near $N^\star$. Set $n(t) = N(t) - N^\star$. Assume that n is small enough and and that f is smooth enough:

$$\frac{dn(t)}{dt} = f\left(N^\star + n(t)\right) \sim n(t)f'\left(N^\star\right) \ .$$

Therefore the behavior of the solution near $N(t) = N^\star$ depends on the sign of $f'(N^\star)$. If $f'(N^\star) < 0$, $N^\star$ attracts the solution $N(t)$: $N^\star$ is a stable steady state. If $f'(N^\star) > 0$, the solution $N(t)$ is ejected: $N^\star$ is an unstable steady state. The global behavior of the solution can be deduced from the study of the stability of the steady state (cf. appendix A.1): the function $N(t)$ is monotonic, and the limit of $N(t)$ is the nearest stable steady state. Several models of this type have been studied: see the exercises.

Let us make a comment on the use of mathematical modeling for predictions. An important question in Ecology deals with the vanishing of a given species. One is tempted then to model the dynamics of a species by a differential equation $\dfrac{dN(t)}{dt} = f(N(t))$ to see whether its solution $N(t)$ can vanish. We immediately see that the answer is negative if $f'(0) > 0$. It means that the conclusion ("the population cannot vanish") is contained in the mathematical model ("$f'(0) > 0$"). This example is very basic, but we always need to keep in mind that making predictions strongly depend on the a priori assumptions done on the model.

2.2.3 Population models with age distribution

When modeling, a first naive idea is to try to obtain a realistic model. By realistic model, we mean a model in which no "real world assumption" has been forgotten. Two objections can be made.

- More realistic models can lead to intricate models! And such realism (of the models) has a cost: we cannot say anything on the models.
- There still remains hidden assumptions that have been forgotten. For instance, make the following experiment in a class. Firstly, study a single population with an autonomous differential equation. The qualitative conclusion is that the size of an isolated population converges to a constant. In other words, there is no oscillation or sophisticated behavior (like chaotic behavior) with one single species. Then ask if some hidden assumptions have been made. Various and interesting answers are obtained, but I have

never had the suggestion that working with continuous-time models instead of discrete-time models has a dramatic influence. Then study the discrete-time logistic model...

We will now study classical demographical models that lead to rather intricate situations. A drawback of the previous demographical models is that the age distribution has not been taken into account. For instance, babies can procreate as soon as they are born! Therefore, assuming delays between birth and procreation seems reasonable.

Mc Kendrick-Von Foerster equation

Let $n(t, a)$ be the population at time t and age range a. The global population at time t is $\int_0^\infty n(t, a)da$. Let $\phi(t, a)$ and $\mu(t, a)$ be the birth and death rates. During an infinitesimal time dt, $\mu(t, a)n(t, a)dt$ people of age a died. The birth rate only influences $n(t, 0)$ (nobody is born with an age $a > 0 \ldots$).

The conservation equation, called Mc Kendrick-Von Foerster equation, is:

$$dn(t, a) = \frac{\partial n}{\partial t}dt + \frac{\partial n}{\partial a}da$$
$$= -\mu(t, a)n(t, a)dt$$

The term $\frac{\partial n}{\partial a}da$ comes from the ageing of the population. Noting that $da/dt \equiv 1$ (After one year, you are one year older!), $n(t, a)$ satisfies the following linear partial differential equation:

$$\frac{\partial n}{\partial t} + \frac{\partial n}{\partial a} = -\mu(t, a)n \ . \tag{2.2}$$

Now we need to specify the boundary conditions. Let $n_0(a)$ be the initial age distribution:

$$n(0, a) = n_0(a) \ . \tag{2.3}$$

The other boundary condition is given by the births:

$$n(t, 0) = \int_0^{+\infty} \phi(t, a)n(t, a)da \ . \tag{2.4}$$

We have taken $+\infty$ as the upper limit of the age for simplicity: of course function $a \to \phi(t, a)$ is a compactly supported function.

Resolution

We indicate a general method for solving the Mc Kendrick-Von Foerster equation when the birth and death rates are independent from the time t: they only depend on the age a.

The operator $n \rightarrow \dfrac{\partial n}{\partial t} + \dfrac{\partial n}{\partial a}$ is a linear first-order operator and there exists a change of variables that transforms the partial differential equation into an ordinary differential equation. Set:

$$\begin{cases} \xi = a \, , \\ \eta = t - a \, . \end{cases}$$

The introduction of the variable η amounts to following a generation through the time. We then have:

$$\frac{\partial n}{\partial \xi} = -\mu(\xi)n \, .$$

This equation is easily solved as:

$$n(\eta, \xi) = f(\eta) \exp\left\{ -\int_0^{\xi} \mu \right\} \, ,$$

where the function f is still unknown. Let us come back to the variables (t, a):

$$n(t, a) = f(t - a) \exp\left\{ -\int_0^{a} \mu \right\} \, . \tag{2.5}$$

We determine the function f for negative values using the initial condition (2.3):

$$n(0, a) = n_0(a)$$
$$= f(-a) \exp\left\{ -\int_0^{a} \mu \right\} \, ,$$

so:

$$f(-a) = n_0(a) \exp\left\{ \int_0^{a} \mu \right\} \, . \tag{2.6}$$

We calculate f for the positive values thanks to relation (2.4) about births. Let:

$$L(a) = \phi(a) \exp\left\{ -\int_0^{a} \mu \right\} \, .$$

The function f, for $t \geq 0$, satisfies:

$$f(t) = \int_0^{\infty} f(t - a)L(a)da \, . \tag{2.7}$$

The homogeneous integral equation (2.7) can be solved using the Laplace transform. We first work as if the Laplace transform of f were defined.

Let:

$$\widehat{f}(\lambda) = \int_0^{+\infty} e^{-\lambda t} f(t) dt \ ,$$

$$\widehat{L}(\lambda) = \int_0^{+\infty} e^{-\lambda t} L(t) dt \ .$$

Then:

$$\widehat{f}(\lambda) = \int_0^{+\infty} L(a) \exp\{-\lambda a\} \int_0^{+\infty} f(t-a) \exp\{-\lambda(t-a)\} dt da$$

$$= \int_0^{+\infty} L(a) \exp\{-\lambda a\} da \int_{-a}^0 f(u) \exp\{-\lambda u\} du$$

$$+ \int_0^{+\infty} L(a) \exp\{-\lambda a\} da \int_0^{+\infty} f(u) \exp\{-\lambda u\} du \ ,$$

and the Laplace transform of f is given by:

$$\widehat{f}(\lambda)(1 - \widehat{L}(\lambda)) = \int_0^{+\infty} L(a) \exp\{-\lambda a\} da \qquad (2.8)$$

$$\int_{-a}^0 n_0(-u) \exp\left\{ \int_0^{-u} \mu \right\} \exp(-\lambda u) du \ .$$

We give some indications on the equation (2.8). Consistently with their biological interpretation, the functions $\phi(a)$ and $\mu(a)$ are compactly supported:

$$L \equiv \int_0^{+\infty} L(t) dt \ < \infty \ .$$

Let us distinguish two cases.

1. $L < 1$. We then have $1 - \widehat{L}(\lambda) > 0$ for $\lambda \geq 0$. The function f is determined by (2.8).
2. $L > 1$. There exists at least one real number λ_0 such that $\lim_{\lambda \to \lambda_0^+} \widehat{f}(\lambda) = +\infty$:

   $f(t)$ is not bounded for $t \geq 0$ and one can expect an explosion of the population. Using the Dominated Convergence Theorem, we choose K such that $\int_0^\infty \exp(-Ka) L(a) da < 1$. Let $g(t) = f(t) \exp(-Kt)$ and $H(a) = L(a) \exp(-Ka)$. Equation (2.7) can be written as follows:

$$g(t) = \int_0^\infty g(t-a) H(a) da \ .$$

$g(t)$ has now a Laplace transform for $\lambda \geq 0$.

By the same computations we obtain:

$$\widehat{g}(\lambda)(1 - \widehat{H}(\lambda)) = \int_0^{+\infty} H(a)\exp\{-\lambda a\}da \qquad (2.9)$$

$$\int_{-a}^0 n_0(-u)\exp\left\{\int_0^{-u}\mu\right\}\exp(-\lambda u)du .$$

We have $1 - \widehat{H}(\lambda) > 0$ for $\lambda \geq 0$ and the function $\widehat{g}(\lambda)$ is determined by (2.9). Let us come back to $f(t) = g(t)\exp(Kt)$. This is an indication -even though not a proof- of an explosion of the population as $t \to +\infty$.

2.3 Pest control: the spruce budworm

The spruce budworm (*Choristoneura fumiferana (Clemens)*) is an insect that damages forests in North America. The spruce budworm lives on and feeds on needles of coniferous trees. Excessive consumption can damage and kill the host. The budworms themselves are preyed primarily by birds, which eat many other insects as well. Our aim is not to make precise qualitative predictions but to see how models can be used to understand the outbreak of the spruce budworm and to evaluate management decisions in the natural resource realm.

The simplest model ([58]) is a single species model, measuring only the spruce budworm population $N(t)$. The idea is the following. If there were not any birds, the spruce budworm population could be described by a logistic model:

$$\frac{dN(t)}{dt} = rN(t)\left(1 - \frac{N(t)}{K}\right) , \qquad (2.10)$$

We add a predation term to take the birds into account. How to model the predation by birds? For large values of $N(t)$, the predation is close to its saturation value. For small population values, as the birds eat other insects, the predation term $p(N)$ rapidly drops to zero. A good candidate for $p(N)$ is a sigmoïdal function, *i.e.*:

$$p(N) = \frac{BN^2}{A^2 + N^2} .$$

The actual differential equation is:

$$\frac{dN(t)}{dt} = rN(t)\left(1 - \frac{N(t)}{K}\right) - \frac{BN(t)^2}{A^2 + N(t)^2} , \qquad (2.11)$$

In order to analyse the model, we express it in non dimensional terms. There are at least two reasons for doing so. Firstly, for the sake of simplicity, we want to reduce the number of relevant parameters. Secondly, as pointed out above, we need to give a precise meaning to "N small". We propose the following change of parameters. Of course, it is not the unique possibility.

$$u = \frac{N}{A} \ , \ \kappa = \frac{K}{A} \ , \ \rho = \frac{rA}{B} \ , \tau = \frac{Bt}{A} \ .$$

A is a parameter greater than 1, and now "u small" means $u \ll 1$. The new equation has only two parameters, ρ and κ:

$$\frac{du}{dt} = \rho u \left(1 - \frac{u}{\kappa}\right) - \frac{u^2}{1 + u^2} \ .$$

$u = 0$ is always a unstable steady state. The other steady states satisfy:

$$f(u; \rho, \kappa) = 0 \ , \tag{2.12}$$

where

$$f(u; \rho, \kappa) = \rho \left(1 - \frac{u}{\kappa}\right) - \frac{u}{1 + u^2} \ .$$

A graphical resolution (*cf.* figure 2.2) clearly indicates that this equation has either one solution u_1, or three, denoted by u_1, u_2 and u_3. We can easily check that u_1 and u_3, when they exist, are stable steady states, and that u_2, when it exists, is an unstable steady state. For given parameters ρ and κ, the function u converges, as $\tau \to +\infty$, to:

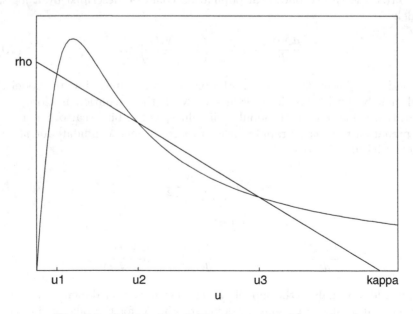

Fig. 2.2. Graphical resolution of equation $f(u; \rho, \kappa) = 0$.

- u_1 if there exists one solution to equation (2.12);
- or
  - u_1 if $u(0) \in (0, u_2)$;
  - u_3 if $u(0) \in (u_2, +\infty)$.

- Outbreak.
  Let the parameter κ be fixed and let's see what happens when ρ varies. Let us give a couple (ρ, κ) such that equation $f(u; \rho, \kappa) = 0$ has three solutions. We start from an initial condition $u(0)$ small. The population converges to the smallest stable steady state u_1. Let us slightly increase the parameter ρ: this corresponds for instance to a decay of the birds predation. Assume that the equation $f(u; \rho, \kappa) = 0$ still has three solutions. The population converges to a new steady state, that we still denote by u_1 since it is qualitatively close to the previous state. Changing ρ has not had any qualitative influence. Let us increase parameter ρ again. As long as equation $f(u; \rho, \kappa) = 0$ has three solutions, the size of spruce budworm population increases only weakly. There is a point when a small change of parameter ρ leads us to the case in which equation $f(u; \rho, \kappa) = 0$ has only one solution. This solution is now qualitatively close to the previous steady state u_3: an outbreak of the models occurs.

- Hysteresis, or "better be safe than sorry".
  Now we start from the outbreak steady state u_3. How can we manage to come back to the refuge steady state u_1? Let us look at the graphical resolution again. A first idea is to reduce parameter ρ until equation $f(u; \rho, \kappa) = 0$ has three solutions again. Unfortunately, we can check that the population size converges to the greatest stable steady state u_3 and not to u_1. The spruce budworm has not been eradicated. Our system is not reversible: this model exhibits a hysteresis effect. To eradicate the spruce budworm, we have to reduce parameter ρ until equation $f(u; \rho, \kappa) = 0$ has one solution. The limiting value ρ can be read on figure 2.2 . Of course this non-reversibility has a cost in terms of environmental management, from both economical and ecological viewpoints.

2.3.1 Specialist and generalist predators

In our study of the spruce budworm dynamics, we introduced the predation term $p(N) = \dfrac{BN(t)^2}{A^2 + N(t)^2}$. Various predation terms may be considered. Usually, predators are classified into two groups.

- A generalist predator eats several kinds of preys. When a prey species vanishes, the predator changes its strategy and prefers to eat another prey species rather than to spend time and energy to hunt a rare prey. A way of modeling a generalist predator is to choose a predation term $p(N)$ such that $p'(0) = 0$.

- A specialist predator can only eat one prey species. When this species vanishes, the predator still continues to hunt this prey. A way of modeling this behavior is to choose a predation term $p(N)$ such that $p'(0) > 0$. An example is given in exercise 2.7.7.

2.4 Interactions in biological systems

2.4.1 Predator-Prey: Lotka-Volterra model

Model

The *historical* predator-prey model is due to Volterra [2]. This model is usually called *Lotka-Volterra model* since it has been simultaneously introduced by Lotka (see [56, 84]). The aim is to explain why one can observe oscillating population sizes[3]. One of our tasks is to understand whether the oscillations are due to some external cause (for instance oscillations of the environment) or are due to the internal dynamics of the species. Let us build the Lotka-Volterra model.

- If there is no predator, the number of preys grows exponentially (Malthusian dynamics);
- If there is no prey, the number of predators vanishes exponentially (Malthusian dynamics);
- The number of deaths among preys is proportional to the number of "meetings" between preys and predators, this number itself is proportional to the product of the sizes of the two species.
- The growth of the size of the predators is proportional to the numbers of deaths among the preys.

Of course these assumptions are very coarse. As usual, the aim is not to build a realistic model, but with few relevant parameters to describe a qualitative behavior, *i.e.* periodic fluctuations. Let $N(t)$ be the number of preys and $P(t)$ the number of predators. Our assumptions lead to the equations:

$$\frac{dN}{dt} = \alpha_1 N - \beta_1 N P , \tag{2.13}$$

$$\frac{dP}{dt} = -\alpha_2 P + \beta_2 N P ,$$

where parameters α_1, α_2, β_1 and β_2 are positive.

---

[2] Vito Volterra, 1860-1940 was interested in mathematical modeling of biological systems after the first world war. The war had considerably reduced fishing in the Adriatic and the relative number of predators with respect to preys had increased. Volterra began a study of analytical models in order to explain such observations.

[3] Several examples of oscillating population sizes can be found in the literature (*e.g.* [67]), the most famous concerning snowshoe hare and Canadian lynx.

Analytic resolution of Lotka-Volterra model

The Lotka-Volterra model can be solved analytically. The steady states of the system (2.13), given by the equations $\left(\dfrac{dN}{dt}, \dfrac{dP}{dt}\right) = (0,0)$, are $(0,0)$ and $(\alpha_2/\beta_2, \alpha_1/\gamma_1)$. The change of variables $x = \beta_2/\alpha_2 \ N$ and $y = \beta_1/\alpha_1 \ P$ transforms the systems as:

$$\frac{dx}{dt} = \alpha_1(1-y)x \ , \tag{2.14}$$

$$\frac{dy}{dt} = -\alpha_2(1-x)y \ . \tag{2.15}$$

The steady states are now $(0,0)$ and $(1,1)$. We eliminate the quadratic terms of the equations. $(2.14)\times\alpha_2 + (2.15)\times\alpha_1$ gives:

$$\alpha_2\frac{dx}{dt} + \alpha_1\frac{dy}{dt} = \alpha_1\alpha_2(x-y) \ ,$$

and $(2.14)\times\alpha_2/x + (2.15)\times\alpha_1/y$ gives:

$$\alpha_2\frac{dx}{xdt} + \alpha_1\frac{dy}{ydt} = \alpha_1\alpha_2(x-y) \ .$$

Time is eliminated:

$$\alpha_2\frac{1-x}{x} \ dx = -\alpha_1\frac{1-y}{y} \ dy \ . \tag{2.16}$$

Set: $C = (x(0)\exp(-x(0)))^{\alpha_2} \ (y(0)\exp(-y(0)))^{\alpha_1}$.
   The variables of (2.16) are separated and the solution is:

$$(x(t)\exp(-x(t)))^{\alpha_2} \ (y(t)\exp(-y(t)))^{\alpha_1} = C \ . \tag{2.17}$$

Now, we need to study the curve defined by (2.17). Let Δ be an arbitrary straight line going through the steady state $(1,1)$. Tedious algebra proves that the intersection of Δ and the curve (2.17) contains two points. The steady point $(1,1)$ is in-between these two intersection points. The trajectories of the Lotka-Volterra system are contained in the curve (2.17), they cannot converge to a steady point (except the trivial solutions $(0,0)$ and $(1,1)$), they are not allowed to make an about-turn inside the curve because of the Cauchy-Lipschitz Theorem: the functions x and y are periodic. Examples are given by figures 2.3 and 2.4.

Average population number

We have seen that solutions to Lotka-Volterra are periodic functions. Let T be the (unknown) period. The average number of preys (resp. predators) is $\dfrac{1}{T}\displaystyle\int_0^T x(t)dt$ (resp. $\dfrac{1}{T}\displaystyle\int_0^T y(t)dt$). An integration of equation (2.14) gives:

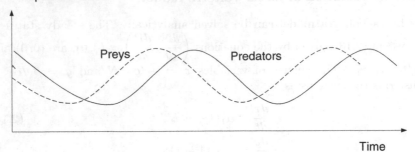

Fig. 2.3. Simultaneous evolution of the numbers of preys and predators

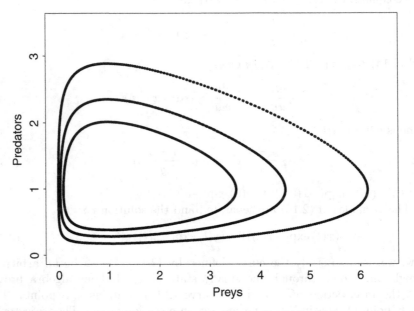

Fig. 2.4. Phase trajectories of the solution to the Lotka-Volterra system

$$x(t) = x(0) \exp \left\{ \alpha_1 \left(t - \int_0^t y(u)du \right) \right\} .$$

Since $x(t + T) \equiv x(t)$:

$$\frac{1}{T} \int_0^T y(t)dt = 1 .$$

Similarly:

$$\frac{1}{T} \int_0^T x(t)dt = 1 .$$

The average number of preys (resp. predators) is $\overline{N} = \dfrac{\alpha_2}{\beta_2}$ (resp. $\overline{P} = \dfrac{\alpha_1}{\beta_1}$).

Super-predator-predator-prey model

Now let us assume that super-predators (*e.g.* human hunters) are hunting both preys and predators. This is a three levels trophic model. Let λ be the number of super-predators and let μ (resp. η) be the way the super-predators hunt the predators (resp. the preys). We generalize the Lotka-Volterra model:

$$\frac{dN}{dt} = (\alpha_1 - \mu\lambda - \beta_1 P)\, N\,,$$

$$\frac{dP}{dt} = (-\alpha_2 - \eta\lambda + \beta_2 N)\, P\,.$$

There are actually two cases.

- $\alpha_1 > \mu\lambda$. This is a Lotka-Volterra model indeed. The average number of preys (resp. predators) is $\dfrac{\alpha_2 + \eta\lambda}{\beta_2}$ (resp. $\dfrac{\alpha_1 - \mu\lambda}{\beta_1}$). As the number of super-predators increases (that is, when λ goes from zero to the critical value α_1/μ), the number of predators decreases, but the number of preys increases. This has been observed, for instance, with rabbits and foxes. The concomitant hunting of rabbits and foxes leads to an explosion of rabbits. This three levels trophic model can be generalized to multi levels trophic models. When the number of trophic levels is odd, the population size of the lowest level is high; when the number of trophic levels is even, the population size of the lowest level is low[4].

- $\alpha_1 < \mu\lambda$. We can easily check that $\dfrac{dN}{N dt} < \alpha_1 - \mu\lambda$: the number of preys decays exponentially. The preys disappear. Similarly, the predators will disappear.

2.4.2 Sketch of a general predator-prey model

The Lotka-Volterra model is sometimes too coarse. It reveals the existence of oscillations in biological systems in a very simple way, but Lotka-Volterra has several drawbacks. Indeed, the periodic solutions of Lotka-Volterra are only determined by the initial conditions which is rather unrealistic. Moreover we have to keep in mind that Lotka-Volterra, contrarily to a common belief, is a marginal two-dimensional dynamical system. It is rather rare for a two-dimensional system to admit an infinity of cycles. Moreover, a lot of two-dimensional systems do not have periodic solutions. A huge number of

---

[4] Plankton is at the lowest level in the trophic level. The abundance of plankton -and therefore the color of the sea- depends on the parity on the number of trophic levels.

predator-prey models (or host-disease models, see exercises 2.7.16 and 2.7.17) has been proposed. We sketch here some outlines of these models (*e.g.* [64, 37]).

$$\frac{dN(t)}{dt} = N(t) \, f(N(t), P(t)) \, ,$$

$$\frac{dP(t)}{dt} = P(t) \, g(N(t), P(t)) \, .$$

The following assumptions on functions f and g can be made.

- The birth rate of the preys is low when the predators are numerous: the function $y \to f(x, y)$ is decreasing.
- The birth rate of preys decreases and the birth rate of predators increases when both predator and prey populations grow: the function $x \to f(rx, rx)$, with r fixed, is a decreasing function and the function $y \to g(ry, ry)$, with r fixed, is an increasing function.
- Prey population is growing when both preys and predators are rare: $f(0, 0) > 0$.
- If there is no more the predator, the number of preys is stable (no Malthusian dynamics) : the function $x \to f(x, 0)$ is negative for x large.
- When the number of predators is large enough, the number of preys cannot grow: the function $y \to f(x, y)$ is negative for y large.
- For a given number of preys, a growth of predators challenges the reproduction of predators: the function $y \to f(x, y)$ is decreasing.
- If there is no more prey, the predators vanish: the function $y \to g(0, y)$ is negative.
- Few predators and many preys contribute to the reproduction of preys: the function $x \to g(x, 0)$ is positive for x large.

The following qualitative conclusions can be drawn from the previous assumptions.

- The point $(0, 0)$ is a steady state. Its stability matrix is $\begin{pmatrix} f(0,0) & 0 \\ 0 & g(0,0) \end{pmatrix}$.
  This is a saddle-point. The extinction of the populations is impossible.
- There is at least one steady state $(N^\star, 0)$ with $N^\star > 0$. Its stability matrix is $\begin{pmatrix} N^\star f_x(N^\star, 0) & N^\star f_y(N^\star, 0) \\ 0 & g(N^\star, 0) \end{pmatrix}$. Since $g(N^\star, 0)$ is non-negative, this point is unstable: in these models, preys cannot live without predators.
- If a steady state $(N^\star, P^\star)$ exists with $N^\star > 0$ and $P^\star > 0$, this is not a saddle-point. Indeed, the product of the eigenvalues of the stability matrix $\begin{pmatrix} N^\star f_x(N^\star, P^\star) & N^\star f_y(N^\star, P^\star) \\ P^\star g_x(N^\star, P^\star) & P^\star g_y(N^\star, P^\star) \end{pmatrix}$ is $N^\star P^\star f_y(N^\star, P^\star) g_x(N^\star, P^\star)$ and is non-negative (consider the behavior of the functions $x \to f(rx, rx)$ and $y \to g(ry, ry)$ for r fixed). This point can either be stable, or unstable.

A graphical reasoning, based on the previous assumptions, shows that there usually exists a connected, bounded domain Ω, with C^1 boundary, contained in $\{x \geq 0, \, y \geq 0\}$ such that:

$$(f(x,y)g(x,y)) \cdot \mathbf{n} > 0$$

$$(x,y) \in \partial\Omega \,,$$

where $\mathbf{n}$ is the normal vector to $\partial\Omega$ at a point (x,y). Then, if the initial condition $(N(0), P(0))$ belongs to Ω, the trajectory remains inside Ω. In other words, there is no demographical explosion and only two qualitative possibilities, according to the Poincaré-Bendixson Theorem[5]. Either the solutions converge to a steady point, or they converge to a limit cycle. That means that a two-species model will either stabilize (steady state case) or will be periodic (limit cycle case). No other qualitative behavior is allowed.

When the functions f and g both depend on a parameter γ, the system may exhibit a *bifurcation* when γ varies. For instance, for a given range of γ, the trajectories may converge to a steady state, though for other values of γ, the trajectories may converge to a limit cycle. Such an example is given in exercise (2.7.12).

2.4.3 Competition and ecological niche

We still assume that two species share the same ecosystem, but now, none of them is a predator of the other. These species are in competition since they use the same single resource. Can these two species coexist or not? This leads to the concept of ecological niche for species. The modeling is similar to the Lotka-Volterra one and is left to the reader. A modeling based on the logistic model is the following:

$$\frac{dN(t)}{dt} = \alpha N(t) \left(1 - \frac{N(t) + \beta_1 P(t)}{K_1}\right) , \tag{2.18}$$

$$\frac{dP(t)}{dt} = \alpha P(t) \left(1 - \frac{P(t) + \beta_2 N(t)}{K_2}\right) .$$

- The birth rates are equal for the two species, but one can check that it does not matter.
- The carrying capacities K_1 and K_2 are supposed to be different.
- The nuisances between the two species are measured by the constants β_1 and β_2. We assume $0 < \beta_1, \beta_2 < 1$.

We do not want to investigate all the cases here. See exercise (2.7.13) for a complete study of the competition models.

The stability matrix at point $(0,0)$ is αId. Steady state $(0,0)$ is still unstable. There are three cases, depending on the value of the parameters K_1, K_2, β_1 and β_2.

---

[5] *cf.* Appendix A.1

- $\beta_2 K_1 > K_2$ and $\beta_1 K_2 < K_1$.

  The stability matrix at the point $(K_1, 0)$ is $\begin{pmatrix} -1 & -\beta_1 \\ 0 & 1 - \frac{\beta_2 K_1}{K_2} \end{pmatrix}$. This point is stable. Similar computations show that $(0, K_2)$ is unstable. There is no other admissible steady state.

- $\beta_1 K_2 > K_1$ and $\beta_2 K_1 < K_2$.

  Similar computations show that $(K_1, 0)$ is unstable and that $(0, K_2)$ is stable.

- $K_2 > \beta_2 K_1$ and $K_1 > \beta_1 K_2$.

  Both $(K_1, 0)$ and $(0, K_2)$ are unstable. Now there exists another admissible steady state:

$$N^\star = \frac{K_1 - \beta_1 K_2}{1 - \beta_1 \beta_2} ,$$

$$P^\star = \frac{K_2 - \beta_2 K_1}{1 - \beta_1 \beta_2} .$$

Its stability matrix $M = (M_{i,j})_{i,j=1,2}$ is given by:

$$M_{1,1} = -\frac{N^\star}{K_1} ,$$

$$M_{1,2} = -\frac{N^\star \beta_1}{K_1} ,$$

$$M_{2,1} = -\frac{P^\star \beta_2}{K_2} ,$$

$$M_{2,2} = -\frac{P^\star}{K_2} .$$

The trace of M is negative and its determinant is positive, this point is stable.

To obtain global qualitative results, graphical results (*e.g.* [8]) are used. For a given point (N, P), the signs of $\frac{dN(t)}{dt}$ and of $\frac{dP(t)}{dt}$ are easily obtained from equation (2.18). We therefore roughly know the direction of the trajectory at this point (N, P). On figures 2.5 and 2.6, a "+" (resp. a "-") has been added to N or P when the function is increasing (resp. decreasing).

- Let us consider the case described by figure 2.5. A trajectory starting from the areas $(P+, N+)$ or $(P-, N-)$ and that remains inside this area is a monotonic function and has a limit. This limit can only be a stable steady state. There is no stable steady state in the area $(P+, N+)$. A trajectory starting from the area $(P+, N+)$ has to go out. The stable steady state $(K_1, 0)$ belongs to the area $(P-, N-)$. A trajectory starting from the area $(P-, N-)$ can either converge to $(K_1, 0)$, or enter in the area $(P-, N+)$. A trajectory starting from the area $(P-, N+)$ cannot go

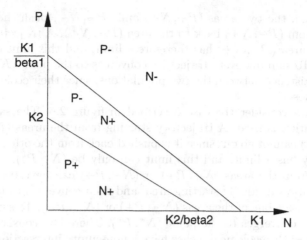

Fig. 2.5. Competition. Case $\beta_2 K_1 > K_2$ and $\beta_1 K_2 < K_1$

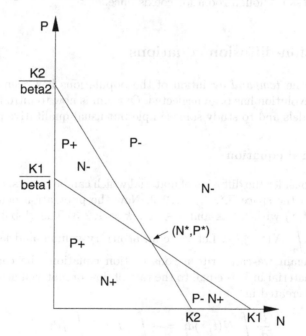

Fig. 2.6. Competition. Case $K_2 > \beta_2 K_1$ and $K_1 > \beta_1 K_2$

out: indeed, the two areas $(P+, N+)$ and $(P-, N-)$ push the trajectory coming from $(P-, N+)$ back to the area $(P-, N+)$. A trajectory starting from the area $(P-, N+)$ has therefore a limit, and this limit can only be $(K_1, 0)$[6] To sum up, every trajectory converges to the point $(K_1, 0)$. There is no coexistence between the two populations, since their ecological niches are too close.

- Let us now consider the case described in figure 2.6. The same graphical resolution is used. A trajectory starting from the areas $(N-, P+)$ or $(N+, P-)$ cannot go out since it is pushed back from the other areas. This trajectory has a limit, and this limit can only be $(N^\star, P^\star)$. A trajectory starting from the areas $(N+, P+)$ or $(N-, P-)$ can have two behaviors: either it stays inside its starting area, and then converges to $(N^\star, P^\star)$, or it enters inside one of the areas $(N-, P+)$ or $(N+, P-)$. To sum up, every trajectory converges to the point $(N^\star, P^\star)$. There is a coexistence of the two species: the ecological niches have a non-empty intersection, but there nevertheless is enough room for coexistence.

2.5 Reaction-diffusion equations

So far, only the *temporal* evolution of the populations has been considered. The *spatial* evolution has been neglected. Our aim is now to introduce *spatial-temporal* models and to study some simple but usual qualitative phenomena.

2.5.1 General equation

Let us now consider the diffusion of material which can be insects, cells, rabbits and so on, in the space $\mathbb{R}^k$, $k = 1, 2, 3$. Now the population density N is a function $N(t, x)$ with $t \in \mathbb{R}$ and $x \in \mathbb{R}^k$, $k = 1, 2, 3$. The global population at time t is $\int_{\mathbb{R}^k} N(t, x)dx$. Let V be an arbitrary volume and let ∂V be its boundary. Again we can write a conservation equation: the change of the amount of material in V is equal to the rate of flow of material across ∂V plus the material created in V:

$$\frac{\partial}{\partial t} \int_V N(t, x)dv = -\int_{\partial V} Jds + \int_V fdv. \tag{2.19}$$

In the most general framework, J and f are functions of N, t and x. Applying the Divergence Theorem to (2.19), the last equation becomes:

$$\int_V \left(\frac{\partial N}{\partial t} + \nabla J - f \right) dv = 0.$$

---

[6] We have admitted that the model is well-posed: there is no negative population size. This is left to the reader.

As the volume V is arbitrary, the integrand must be zero:

$$\frac{\partial N}{\partial t} + \nabla J = f \; . \tag{2.20}$$

In the absence of external factors, a flow proportional to ∇N is the simplest choice[7]:

$$J = -D\nabla N \; .$$

In the most general situation, D is itself a function of N and x. If we assume D to be constant, equation (2.20) becomes:

$$\frac{\partial N}{\partial t} = f + \Delta N \; . \tag{2.21}$$

Equation (2.21) [8] is not sufficient. We need to specify an initial condition $N(0, x)$ and a boundary condition.

An historical reaction-diffusion equation like (2.21) is the so-called Fisher equation[9] in which the birth/death function f is logistic:

$$\frac{\partial N}{\partial t} = rN\left(1 - \frac{N}{K}\right) + \Delta N \; . \tag{2.22}$$

2.5.2 Solution control: maximum principle

When writing a reaction-diffusion equation, the first two questions are the following:

- Is the model well-posed? In other words, does there exist solutions, are they positive?
- Is an explosion of the population possible?

Consider the reaction-diffusion with a vanishing boundary condition (Dirichlet boundary condition):

$$\frac{\partial N}{\partial t} = f(N) + \Delta N \; ,$$
$$N(t, x) = 0 \; , \; x \in \partial \Omega \; ,$$
$$u(0, x) = \phi(x) \; ,$$

with $t \in [0, T]$ and $x \in \Omega$. ϕ is the initial population and Ω is assumed to be compact. Moreover we assume that f is Lipschitz:

---

[7] An analogy with atmospheric pressure can be done.

[8] See appendix A.2 for theoretical results on partial differential equations.

[9] Ronald Fisher, 1890-1962, is mostly known for his statistical works, but he has also worked on genetical and population dynamical models.

$$f(x) - f(y) \geq -C|x - y| .$$

Suppose that we know a sub-solution N^1 and a super-solution N^2. N^1 and N^2 are two functions on $[0, T] \times \Omega$ that satisfy:

$$\frac{\partial N^1}{\partial t} - \Delta N^1 \leq f(N^1) ,$$

$$\frac{\partial N^2}{\partial t} - \Delta N^2 \geq f(N^2) ,$$

$$N^2 \geq N^1 \text{ on } \partial\Omega ,$$
$$N^2(0, x) \geq N^1(0, x) .$$

Let $Z = e^{-(C+1)t}(N^2 - N^1)$, where C is the Lipschitz constant of f. If N^1 and N^2 are bounded functions, it is rather easy to obtain a constant C. Now suppose that N^1 and N^2 are twice continuously differentiable[10]. A little bit of algebra proves that, if Z is negative:

$$\frac{\partial Z}{\partial t} - \Delta Z \geq -Z . \tag{2.23}$$

A minimum of function Z satisfy $\dfrac{\partial Z}{\partial t} - \Delta Z \leq 0$. Inequality (2.23) proves that this minimum cannot be reached for a negative value of Z. It follows that $N^2 - N^1$ is a positive function.

Let us apply this result to Fisher equation (2.22). Assume that the initial population is less than the carrying capacity K, i.e. $0 \leq \phi \leq K$. Let N be the solution. Function $N_0 \equiv 0$ is a sub-solution and function $N_K \equiv K$ is a super-solution. Solution N can be considered both as a sub-solution and super-solution. An application of the maximum principle to the couples $(N^1, N^2) = (N_0, N)$ and $(N^1, N^2) = (N, N_K)$ shows that $0 \leq N \leq K$. Fisher equation is well-posed and there is no explosion of the population.

2.5.3 Steady solution: stability

A steady solution is a solution that does not evolve anymore.

Definition 2.5.1 *Steady solution.*
A one-variable function $N_0(x)$ is a steady solution of the reaction-diffusion equation (2.21) if

$$\Delta N_0 + f(N_0) = 0 .$$

---

[10] Without this differentiability condition, the proof is more intricate. See section A.2.5 for more general statements.

See [28, Ch.5] for a rigorous presentation of the stability of a steady solution: we will only give here an heuristic idea of the stability of a steady solution.

Let $\varepsilon(x)$ be a small perturbation. Let $N(t,x)$ be the solution to the reaction-diffusion equation (2.21) with initial condition $N(0,x) = N_0(x) + \varepsilon(x)$. N_0 is a stable steady solution if $N(t,x)$ converges to N_0 uniformly with respect to x as $t \to \infty$.

Suppose that f is linearizable about N_0, i.e. there exists a linear operator L such that:

$$f(N_0 + \varepsilon) \sim f(N_0) + L\,\varepsilon\,.$$

Set $\varepsilon(t,x) = N(t,x) - N_0(x)$. A linearization of the reaction-diffusion equation (2.21) leads to:

$$\frac{\partial \varepsilon}{\partial t} \sim (L + \Delta)\,\varepsilon\,.$$

If there exists $\beta > 0$ such that the real part of the eigenvalues of the operator $L + \Delta$ are less than $-\beta$, then the solution N_0 is a steady solution.

Let us give an example. The function $N(t,x) \equiv K$ is a steady solution of the one-dimensional Fisher equation (2.22). Set $u(t,x) = N(t,x)/K$. The Fisher equation becomes:

$$\frac{\partial u}{\partial t} = ru(1 - u)\,,$$

$$u(0,0) = 1\,,$$
$$u(0,1) = 1\,.$$

Solution $u(t,x) \equiv 1$ is of course a steady solution. A linearization about $u(t,x)$ leads to:

$$\frac{\partial \varepsilon}{\partial t} = -r\varepsilon + \frac{\partial^2 \varepsilon}{\partial x^2}\,,$$
$$\varepsilon(0,0) = 0\,,$$
$$\varepsilon(0,1) = 0\,.$$

The eigenvalues of the operator $\varepsilon \to -r\varepsilon + \dfrac{\partial^2 \varepsilon}{\partial x^2}$ are $-r - (\pi/2 + 2k\pi)^2$, with k integer. The steady solution 1 is stable.

2.5.4 The propagation of the Larch Bud Moth

The caterpillars of the Larch Bud Moths (*Zeiraphera diniana*) damage the larches. Sexual dimorphism is marked at the adult stage: females are obese and therefore less moving; on the other hand, males are good flyers. So, this species is poorly moving from a population dynamics point of view, but it is considerably more movable from a genetic population point of view.

The Larch Bud Moths have been observed since the beginning of the XIXth century in Engiadina (Switzerland). It has been systematically studied ([4]) from 1949 to 1976 in five areas covering around 630 km among the Alpine arc: Lungau (Austria), Val Aurina (Italy), Engiadina (Switzerland), Goms (Switzerland) and Névache (France). Measurements of population have been done by counting the number of caterpillars per Bud branch: we must keep in mind that the real data are not error-free! Experimental data are summed up in Figures 2.7 and 2.8. Two phenomena clearly appear.

- Spatial oscillations.
  Let us have a look on a given area (*cf.* figure 2.8). The dynamics of the Moths is periodic, with a succession of highly elevated peaks, corresponding to the crisis of the pest, and bottoms, corresponding to quiet periods. The period between two peaks is around nine years. This typically is the dynamics of an host-disease model: these models are mathematically close to predator-prey models we have seen before.
- Propagation through time.
  A propagation phenomenon clearly appears when looking to the spatio-temporal repartition: the damages move slightly forward from West to East, with a three-year period. Let us model the Alpine arc by a one-dimensional curve. The spatial dynamics of the Moth looks like a wave traveling along this curve. The careful study of [4] indicates that this traveling wave is not due to some external factor, but is due to the internal dynamics of the Moth. Now our aim will be to investigate a reaction-diffusion equation, like Fisher equation, and to see whether these equations generate traveling waves.

2.5.5 Propagation

The aim of this section is to explain the observations and the traveling waves. Let us consider the basic waves: the waves evolve without deformation and with constant speed. In other words, we are looking for functions $N(t, x)$ such that $N(t, x) \equiv N(t', x')$, as soon as $x' - x = c(t' - t)$, for a given wave speed c.

One-dimensional traveling waves

We will restrict ourselves to the one-dimensional case: the spatial variable x is a one-dimensional variable. Take $N(t, x) = N(x - ct) = \mathcal{N}(z)$, where $z = x - ct$ is a wave variable. Denote by $\mathcal{N}'$ and $\mathcal{N}''$ the first and second derivatives of $\mathcal{N}(z)$. The function $\mathcal{N}(t, x)$ is the solution to the reaction-diffusion

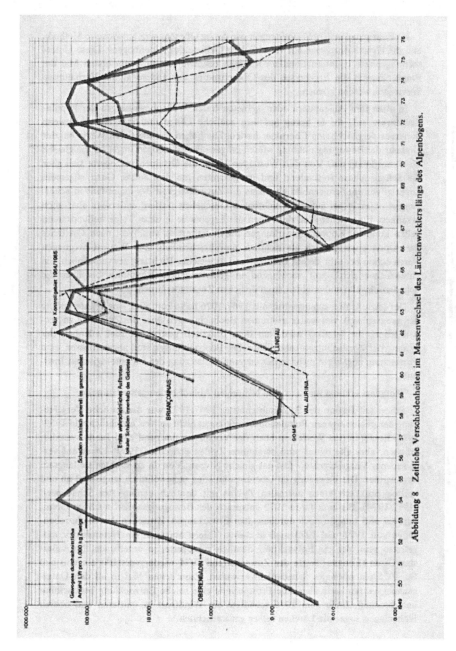

Abbildung 8   Zeitliche Verschiedenheiten im Massenwechsel des Lärchenwicklers längs des Alpenbogens.

Fig. 2.7. Spatio-temporal evolution of the Larch Bud Moths (logarithmic scale)

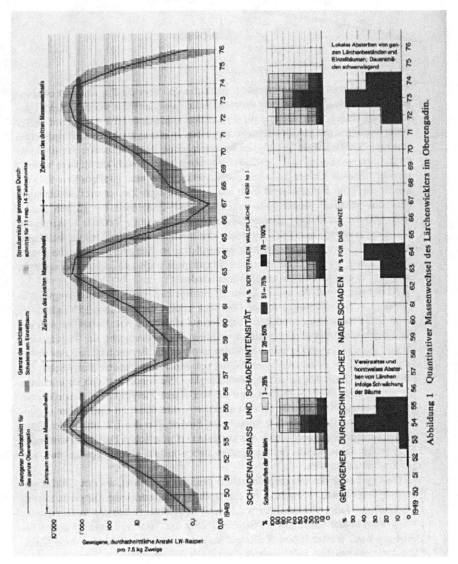

Fig. 2.8. Temporal evolution of the Larch Bud Moths in Engiadina (logarithmic scale)

equation (2.21). Substituting the traveling wave $\mathcal{N}$ into (2.21), $\mathcal{N}$ satisfies the ordinary differential equation:

$$\mathcal{N}'' + c\mathcal{N}' + f(\mathcal{N}) = 0 , \tag{2.24}$$

We will study the steady state of equation (2.24). This gives qualitative results on the traveling waves and on the wave speeds.

Set:

$$U = \mathcal{N} ,$$
$$V = \mathcal{N}' .$$

Equation (2.24) becomes:

$$U' = V , \tag{2.25}$$
$$V' = -cV - f(U) .$$

A steady state of (2.25) is a couple $(U^\star, 0)$ where $U^\star$ satisfies $f(U^\star) = 0$. Let us linearize equation (2.25) about $(U^\star, 0)$. The eigenvalues satisfy:

$$\lambda^2 + c\lambda + f'(U^\star) = 0 .$$

We will then study these steady states as usual.

For biological reasons[11], function f vanishes at 0: $f(0) = 0$. $(0,0)$ is a steady state. When $c^2 < 4f'(0)$, point $(0,0)$ is a spiral point. The traveling wave crosses the half-plan $U < 0$. The population is not allowed to be negative, these solutions have to be rejected. There is a minimal wave speed.

When functions U and V are bounded, we know from Poincaré-Bendixson Theorem that the trajectories of (2.25) converge either to a stable steady point, or to a limit cycle. We will see that the limit cycle is forbidden. Let us introduce the so-called Lyapunov function $\mathcal{L}$:

$$\mathcal{L} = \frac{1}{2}\mathcal{N}'^2 + F(\mathcal{N})$$
$$= \frac{1}{2}V^2 + F(U) ,$$

with

$$F(x) = \int_0^x f(u)du .$$

We will not make an intensive use of Lyapunov function here. An analogy with mechanics can be done. The Lyapunov function can be viewed as the energy of the system. A dissipative system has a decreasing energy and cannot have any limit cycle. We can easily check that $\mathcal{L}' = -c\mathcal{N}'^2$. Function $\mathcal{L}$ is monotonic. The trajectory cannot converge to a limit cycle. To sum up, either the trajectory of (2.25) converges to a stable steady point, or it explodes.

---

[11] There is no spontaneous generation.

Traveling waves for the one-dimensional Fisher equation

Let us recall that the Fisher equation is:

$$\frac{\partial N}{\partial t} = rN\left(1 - \frac{N}{K}\right) + \frac{\partial^2 N}{\partial x^2}\ .$$

A traveling wave $\mathcal{N}$, with $z = x - ct$ and $\mathcal{N}(z) = N(t,x)$ satisfies:

$$\mathcal{N}'' + c\,\mathcal{N}' + r\mathcal{N}\left(1 - \frac{\mathcal{N}}{K}\right) = 0. \tag{2.26}$$

Set $\mathcal{U}(z) = \mathcal{N}(z)/K$, (2.26) becomes:

$$\mathcal{U}'' + c\,\mathcal{U}' + r\mathcal{U}(1 - \mathcal{U}) = 0\ .$$

The steady points are $(0,0)$ and $(1,0)$.

- Point $(0,0)$. We reject the case $c^2 < 4r$: it generates a stable spiral point.
  When $c^2 > 4r$, the eigenvalues are $\lambda_\pm = \dfrac{-c \pm \sqrt{c^2 - 4r}}{2}$ and are negative.
  Point $(0,0)$ is a stable node.

- Point $(1,0)$. The eigenvalues are $\lambda_\pm = \dfrac{-c \pm \sqrt{c^2 + 4r}}{2}$. λ_+ is positive, λ_- is negative. Point $(1,0)$ is a saddle point.

The reader can check that there exists an analytical solution, starting from 1 and converging to 0, with speed $c = r\dfrac{5}{\sqrt{6}}$, and given by:

$$\mathcal{U}(z) = \left(1 + (\sqrt{2} - 1)\exp\left(\frac{z}{\sqrt{6}}\right)\right)^{-2}.$$

According to the qualitative considerations - stability of the steady points, no limit cycles, analytical solution - we can roughly draw the traveling waves of the Fisher equation (*cf.* figure 2.9, see Exercise 2.7.19).

We should wonder whether these traveling waves can really be observed in "real life". Let us have an analogy with the steady point. If a steady point is unstable, any perturbation will remove the trajectory from this point: this point cannot be observed in "real life". On the other hand, a small perturbation has no real influence on a stable steady point. This is the same for a traveling wave: if it is unstable, we cannot observe this solution in "real life". The question of stability of traveling waves is a difficult one. Let us briefly sketch a rough way of studying this stability. (we refer to [1, 7, 49, 78] for more general results). Assume that $N^\star$ is a solution of:

$$\frac{\partial N}{\partial t} = f(N) + \Delta N\ .$$

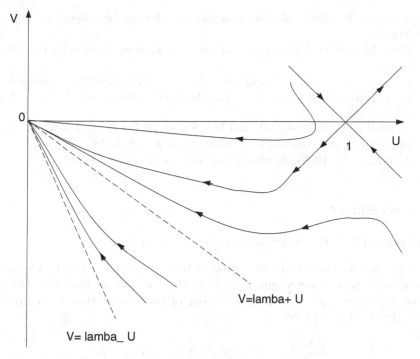

Fig. 2.9. Qualitative drawing of the traveling waves of the Fisher equation

Let ε be a small perturbation. We want $N^\star + \varepsilon$ to be the solution to the same equation. If the linearization is valid, the equation becomes:

$$\frac{\partial \varepsilon}{\partial t} = \varepsilon f'(N^\star) + \Delta \varepsilon .$$

The maximum principle can then be used to control the solution to this equation and therefore to decide whether the traveling wave is stable or not.

2.6 Bibliography

- AUER, C. (1977). Dynamik von Laerchenwickler-Populationen laengs des Alpenbogens (in German). *Eidg. Anst. fuer Forstl. Versuchsweses.* **53** 71-105. Fasc. 2.
- CUSHING, J. (1977). *Integrodifferential Equations and Delay Models in Population Dynamic.* Lecture Notes in Biomathematics, Springer-Verlag.
- FIFE, P. (1979). *Mathematical Aspects of Reacting and Diffusing Systems.* volume 28. Lecture Notes in Biomathematics, Springer-Verlag.
- GOLDBETTER, A. (1996). *Biochemical Oscillations and Cellular Rhythms: the Molecular Bases of Periodic and Chaotic Behavior.* Cambridge University Press.

- JOLIVET, E. (1983). *Introduction aux modèles mathématiques en biologie.* Masson.
- KOT, M. (2001). *Elements of Mathematical Ecology.* Cambridge University Press.
- LUDWIG, D., JONES, D. and HOLLING, C. (1978). Qualitative analysis of insect outbreak systems: the spruce budworm and forest. *J. Anim. Ecol.* **47** 315-332.
- LUDWIG, D., ARONSON,D. and WEINBERGER,H (1979). Spatial patterning of the spruce budworm, *J. Math. Biology* **8** 217-258.
- MURRAY, J. (1990). *Mathematical Biology.* Springer-Verlag.

2.7 Exercises

Exercise 2.7.1 *The growth of bacteria in a Petri dish.*

Let us consider the growth of a mass of bacteria in a Petri dish. The mass grows uniformly in every direction. Only the bacteria on the surface of the mass reproduce. Let $N(t)$ be the number of bacteria at time t. Justify the following model and solve it:

$$\frac{dN(t)}{dt} = r\, N^{2/3}(t)\,, \quad N(0) > 0\,, \; r > 0\,.$$

Exercise 2.7.2 *Gompertz model.*

Study the one species model given by:

$$\frac{dN(t)}{dt} = -\alpha N(t) \log\left(\frac{N(t)}{K}\right)\,,$$

with $\alpha, K > 0$ and $N(0) \geq 0$.

Exercise 2.7.3 *Demographic models.*

Let us consider the following model:

$$\frac{dN(t)}{dt} = a\frac{N(t)}{\alpha}\left(1 - \left(\frac{N(t)}{K}\right)^{\alpha}\right)\,,$$

where a and α are positive.

1. Recognize this model when $\alpha = 1$ and $\alpha \to 0$.
2. What is the limit of $N(t)$ when $t \to \infty$?

Exercise 2.7.4 *Seasonal capacity model.*

A species N is subject to a seasonal periodic constraint that changes its carrying capacity. The proposed model is:

$$\frac{dN(t)}{dt} = \alpha N(t) \left(1 - N(t)\frac{1 + \beta \cos(\gamma t)}{K} \right) ,$$

where $\alpha > 0$, $K > 0$, $\gamma > 0$ and $0 < \beta < 1$.

1. Solve the equation (Hint: set $y = 1/N$).
2. Compare with the behavior of the logistic model:

$$\frac{dN(t)}{dt} = \alpha N(t) \left(1 - \frac{N(t)}{K} \right) .$$

Exercise 2.7.5 *Constant-rate harvesting.*

Study the logistic model with constant-rate harvesting:

$$\frac{dN(t)}{dt} = \alpha N(t) \left(1 - \frac{N(t)}{K} \right) - h ,$$

where $\alpha > 0$, $K > 0$, $h > 0$.

Exercise 2.7.6 *Optimal harvesting.*

Consider the logistic model with proportional-rate harvesting:

$$\frac{dN(t)}{dt} = \alpha N(t) \left(1 - \frac{N(t)}{K} \right) - hN(t) ,$$

where $\alpha > 0$, $K > 0$, $h > 0$. For $T > 0$, let $Y_T(h)$ be the total yield:

$$Y_T(h) = \int_0^T hN(t)dt .$$

Compute the optimal harvesting rate h_T that maximizes $Y_T(h)$. What is $\lim_{T \to +\infty} h_T$?

Exercise 2.7.7 *Generalist versus specialist predator.*

Consider the one-single species model with predation:

$$\frac{dN(t)}{dt} = rN(t) \left(1 - \frac{N(t)}{K} \right) - p(N(t)) ,$$

where all the parameters are positive. Study the model (population outbreak, hysteresis) with the following predation terms:

$$p(N) = B \left(1 - \exp \left(-\frac{N^2}{A^2} \right) \right) ,$$

$$p(N) = \frac{BN}{A + N} .$$

Exercise 2.7.8 *Insect pest control.*

Consider an insect population $N(t)$. A method to eradicate this population consists in introducing sterile insects $n(t)$. These sterile insects prejudice the reproduction of the other insects. The following model is proposed:

$$\frac{dN(t)}{dt} = \left[\frac{aN(t)}{N(t) + n(t)} - b\right] N(t) - kN(t)(N(t) + n(t)) \; ,$$

$$N(0) > 0 \, , n(0) > 0 \; .$$

The parameters a, b and k are positive.

1. We assume that the population of sterile insects is constant $n(t) \equiv n$. Give a condition on n to be sure of the eradication of $N(t)$.
2. We assume that the sterile insects are dropped once and that the dynamics of the sterile insects is Malthusian:

$$\frac{dn(t)}{dt} = -bn(t) \; .$$

Is the eradication of $N(t)$ sure?
3. We assume that a part $\gamma > 0$ of the insects are born sterile:

$$\frac{dn(t)}{dt} = \gamma N(t) - bn(t) \; .$$

Give a condition on γ such that the only steady state is $(0,0)$.

Exercise 2.7.9 *Particular solution of Mc Kendrick-Von Foerster equation.*

Let the birth and death rates be constant. Find initial condition $n_0(a)$ such that $n(t, a) = \alpha \exp(\beta t) \exp(-\gamma a)$ is a solution of Mc Kendrick-Von Foerster equation. Is this solution realistic?

Exercise 2.7.10 *Logistic predator-prey model.*

Study the steady state and their stability in the model:

$$\frac{dN(t)}{dt} = \alpha_1 N(t) \left(1 - \frac{N(t)}{K} - \beta P(t)\right) \; ,$$

$$\frac{dP(t)}{dt} = -\alpha_2 P(t)(1 - \gamma N(t)) \; ,$$

where all the parameters are positive.

Exercise 2.7.11 *Application of the Poincaré-Bendixson Theorem.*

Show the existence of a limit cycle, for $a > 0$, in the equation:

$$\frac{dx(t)}{dt} = ax(t) - by(t) - x(t)(x^2(t) + y^2(t)) \; ,$$

$$\frac{dy(t)}{dt} = bx(t) + ay(t) - y(t)(x^2(t) + y^2(t)) \; .$$

Exercise 2.7.12 *Bifurcation in a predator-prey model.*

Consider the predator-prey model:

$$\frac{dN(t)}{dt} = N(t)\left\{r\left(1 - \frac{N(t)}{K}\right) - \frac{aP(t)}{b + N(t)}\right\},$$

$$\frac{dP(t)}{dt} = P(t)\left\{c\left(1 - d\frac{P(t)}{N(t)}\right)\right\},$$

where all the parameters are positive and $N(0) > 0$, $P(0) > 0$.

1. Propose a change of variables such that the new model becomes:

$$\frac{du}{d\tau} = u(1 - u) - \alpha\frac{uv}{u + \beta},$$

$$\frac{dv}{d\tau} = \gamma v\left(1 - \frac{v}{u}\right).$$

2. Show that there exists an unique steady state $(u^\star, v^\star)$ in the quadrant $u > 0$, $v > 0$.
3. Find the conditions on α, β, γ such that this steady state is stable.
4. What happens when the point $(u^\star, v^\star)$ is unstable?

Exercise 2.7.13 *Competition model (ctd.).*

This exercise is an extension of the competition model (*cf.* section "Competition and ecological niche"):

$$\frac{dN(t)}{dt} = \alpha_1 N(t)\left(1 - \frac{N(t) + \beta_1 P(t)}{K_1}\right),$$

$$\frac{dP(t)}{dt} = \alpha_2 P(t)\left(1 - \frac{P(t) + \beta_2 N(t)}{K_2}\right),$$

where all the parameters are positive.

1. Find the steady states and their stability.
2. Use a graphical resolution in order to find the global behavior of the system.
3. What is the actual influence of the parameters α_1 and α_2?

Exercise 2.7.14 *Another competition model.*

Study analytically the following competition model.

$$\frac{dN(t)}{dt} = (\alpha_1 - \gamma_1(\beta_1 N(t) + \beta_2 P(t)))N(t),$$

$$\frac{dP(t)}{dt} = (\alpha_2 - \gamma_2(\beta_1 N(t) + \beta_2 P(t)))P(t).$$

(Hint: eliminate the quadratic terms.)

Exercise 2.7.15 *Mutualism.*

The interaction between two species is not always a competition or a predation interaction; the interaction between two species could be to the advantage of both species. So is mutualism.

1. The simplest model is:

$$\frac{dN(t)}{dt} = \alpha_1 N(t)(1 + \beta_1 P(t)) ,$$

$$\frac{dP(t)}{dt} = \alpha_2 P(t)(1 + \beta_2 N(t)) ,$$

   where all the parameters are positive. Study the limit of $(N(t), P(t))$ as $t \to +\infty$. Is this model realistic?
2. We incorporate carrying capacities for both species:

$$\frac{dN(t)}{dt} = \alpha_1 N(t) \left(1 - \frac{N(t) - \beta_1 P(t)}{K_1} \right) ,$$

$$\frac{dP(t)}{dt} = \alpha_2 P(t) \left(1 - \frac{P(t) - \beta_2 N(t)}{K_2} \right) ,$$

   where all the parameters are positive. Show that the behavior of $(N(t), P(t))$ as $t \to +\infty$ depends on the sign of $1 - \beta_1 \beta_2$.

Exercise 2.7.16 *Rabies pest.*

In 1979, a rabies pest, coming from East Europe, arrived in France from the east border. Foxes were the main vehicle of the rabies.

1. We consider the foxes to be divided into two groups, infective I and susceptible S. The proposed model is:

$$\frac{dS(t)}{dt} = r(S(t) + I(t)) \left(1 - \frac{S(t)}{K} \right) - \beta S(t)I(t) ,$$

$$\frac{dI(t)}{dt} = \beta S(t)I(t) - uI(t) ,$$

   where all the parameters are positive.
   a) Justify this model.
   b) Find the stability of the steady state $(S, I) = (K, 0)$, give conditions on the parameters that allow an outbreak of the pest. Discuss these conditions. You can adopt the point of view of the rabies virus! What is your best strategy?
2. A first method to eradicate the rabies pest consist in killing the foxes, for instance by giving a bonus to the hunters. This method is modeled by:

$$\frac{dS(t)}{dt} = r(S(t) + I(t))\left(1 - \frac{S(t)}{K}\right) - \beta S(t)I(t) - cS(t) ,$$

$$\frac{dI(t)}{dt} = \beta S(t)I(t) - uI(t) - cI(t) ,$$

where c is the hunting parameter: hunting is done both on infective and susceptible. Give conditions on the parameters that prevent the rabies pest.

3. A second method to eradicate the rabies pest consist in vaccinating the foxes. This method is modeled by:

$$\frac{dS(t)}{dt} = r(S(t) + I(t))\left(1 - \frac{S(t)}{K}\right) - \beta(1 - v)S(t)I(t) ,$$

$$\frac{dI(t)}{dt} = \beta(1 - v)S(t)I(t) - uI(t) ,$$

where v is the vaccinating parameter. Give conditions on the parameters that prevent the rabies pest.

4. Which method do you chose?[12]

Exercise 2.7.17 *Epidemiology model (from [45, 46, 47]).*

Consider a disease that, after recovery, confers immunity. The population at time t, denoted by $N(t)$, is divided into three groups: the susceptible S, the infective I and the removed group R. The global population is then $N(t) = S(t) + I(t) + R(t)$. The model dynamics is then:

$$\frac{dS(t)}{dt} = -rS(t)I(t) ,$$

$$\frac{dI(t)}{dt} = rS(t)I(t) - aI(t) ,$$

$$\frac{dR(t)}{dt} = aI(t) ,$$

with initial populations $S(0)$, $I(0)$ positive and $R(0) = 0$. The parameters r and a are positive.

1. Justify this model.
2. Show that $N(t)$ is constant.
3. Show that the functions S, I and R are positive.
4. Show that if $S(0) < \dfrac{a}{r}$, then $I(t)$ is a decreasing function. What is its limit as $t \to +\infty$?

---

[12] The chosen method was the oral vaccination of the foxes. Vaccine-impregnated baits that looked like meatballs were dropped by airplane.

5. We assume that $S(0) > \dfrac{a}{r}$. Let $I_{\sup} = \sup\limits_{t \geq 0} I(t)$.

   a) Show that there exists $t_0 > 0$ such that $I(t_0) = I_{\sup}$.

   b) Calculate $\dfrac{dI(t)}{dt} + \dfrac{dS(t)}{dt} - \dfrac{a}{rS(t)}\dfrac{dS(t)}{dt}$.

   c) Calculate $I_{\sup}$ in terms of $I(0)$, $S(0)$, a and r.
6. Give a qualitative interpretation of the previous results.

Exercise 2.7.18 *Spatial spread of an epidemic.*

The population consists of two populations, infectives $I(x,t)$ and susceptibles $S(x,t)$, $x \in \mathbb{R}$, $t \geq 0$.

1. Justify the following dispersion model:

$$\frac{\partial}{\partial t}S(x,t) = -rS(x,t)I(x,t) + \frac{\partial^2}{\partial x^2}S(x,t) ,$$

$$\frac{\partial}{\partial t}I(x,t) = rS(x,t)I(x,t) - aI(x,t) + \frac{\partial^2}{\partial x^2}I(x,t) .$$

2. Study the travelling waves solutions with boundary conditions:

$$\lim_{x \to \pm\infty} I(x,t) = 0 .$$

   Show that there exists a minimal wave speed. Is such a travelling wave allowed by the Fisher equation?

Exercise 2.7.19 *Travelling waves of Fisher equation.*

Consider the differential equation:

$$\frac{d^2q}{dt} + c\frac{dq}{dt} + f(q) = 0 ,$$

with $f(0) = f(1) = 0$, $f'(0) > 0$, $f'(1) < 0$ and $f''(q) < 0$ for all $q \in [0,1]$.
   Let:

$$c \geq 2\sqrt{f'(0)} ,$$

$$\alpha = -\frac{1}{f'(1)}\left(-c/2 + \sqrt{c^2/4 - f'(1)}\right) ,$$

$$\beta = \frac{1}{f'(0)}\left(c/2 - \sqrt{c^2/4 - f'(0)}\right) .$$

Define the domain D:

$$D = \{(p,q) \text{ s.t. } 0 \leq q \leq 1,\ -\beta f(q) \leq p \leq -\alpha f(q)\} .$$

1. Check that $\beta > \alpha > 0$.
2. Show that D is invariant under the flow:

$$\frac{dq}{dt} = p,$$

$$\frac{dp}{dt} = -cp - f(q).$$

3. Show that D contains the unstable manifold of the point $(0, 1)$.
4. Conclude and apply to the Fisher equation.

Exercise 2.7.20 *Spatial repartition of the spruce budworm (from [57]).*

Consider the one-dimensional spatio-temporal model of the spruce budworm:

$$\frac{\partial}{\partial t} u(x, t) = f(u(x, t)) + \frac{\partial^2}{\partial x^2} u(x, t),$$

with

$$f(u) = \rho u \left(1 - \frac{u}{\kappa}\right) - \frac{u^2}{1 + u^2}.$$

Study qualitatively the travelling waves of the model.

Exercise 2.7.21 *Advection-reaction-diffusion equation.*

The evolution of a population $N(t, x)$, $t > 0$, $x \in \mathbb{R}$ is given by the following equation:

$$\frac{\partial N(t, x)}{\partial t} = D \frac{\partial^2 N(t, x)}{\partial x^2} - k \frac{\partial N(t, x)}{\partial x} + a N(t, x),$$

the parameters D, k and a are positive, the initial population is M and is concentrated in 0.

1. Justify this model.
2. Solve the equation (Hint: take the Fourier transform of $N(t, x)$).
3. Study the level curve $N(t, x(t)) = ct$ when $t \to \infty$ and find the asymptotical propagation speed of the population spreading.

Exercise 2.7.22 *Delay logistic model.*

1. Consider the following model:

$$\frac{dN(t)}{dt} = N(t)(1 - N(t - T))$$

where T is a positive delay constant and with initial condition:

$$N(t) = \phi(t), \quad t \in [0, T].$$

Justify this delay model from a biological point of view.

2. Assume that the linearization $n(t) = N(t) - 1$ about 1 is valid. The new equation becomes then:

$$\frac{dn(t)}{dt} = -n(t - T) \,. \tag{2.27}$$

3. Show that there exists an infinity of solutions to (2.27) of the form $n(t) = ce^{\lambda t}$, with c being a real number and λ be a complex number. (Hint: show that 0 is an essential singularity of the complex valued function $z \to e^{1/z}$ and then use the Picard Theorem.)
4. What do you think about the stability of point 1?

3

Discrete-time dynamical systems

3.1 Introduction

In the previous section, devoted to continuous-time dynamical systems, we regarded time as a continuous variable. In some biological situations, such hypothesis is not relevant. For instance, we can think of the reproduction of some animals or plants which only occur during a short period in the year. In such cases, it will be more relevant to think in terms of discrete time, the time step being equal to a year. The writing of a discrete-time dynamical model does not put any major problem. In fact there is a kind of heuristic parallel between continuous and discrete-time models. Thus a time wise demographical population will be modeled either with an autonomous differential equation or with a recurrent equation; a population considering age and sexual maturity will be modeled with a Mc Kendrick-Von Foerster-type partial derivative equation or with a discrete delay equation, ... We could naively believe that continuous-time and discrete-time models have the same qualitative characteristics. We will see that it is not the case and that qualitative models, which seem heuristically close to the continuous models we studied previously, display drastically different behaviors. This will constitute the core of this chapter. We will nevertheless start with some basic reminders on discrete recurrent equations with or without delay, but we will not linger.

Before dealing with the study of discrete-time models, let us start with a short reminder of the qualitative characteristics of the continuous-time dynamical systems presented in the previous chapter. In our study of dynamical systems in modeling an isolated population or two populations in interaction, we saw that only three qualitative behaviors were possible because of the Poincaré-Bendixson Theorem:

- convergence to a steady point,
- convergence to a limit cycle,
- blow-up of the number of people (this type of model is not relevant because of it is unrealistic in the long run).

Let us emphasize, once again that in one dimension (that of an isolated population), a continuous-time dynamical system cannot show an oscillating behavior. Such behaviors are very stable: except for an hypothetical preliminary behavior, the system's course quickly stabilize around a steady point or around a limit cycle. Except for hypothetically critical values, an alteration of the initial characteristics of the system does not have an impact of the resulting characteristics.

We will focus on the case of an isolated population, taking its temporal evolution into account. Our study-case will be the discrete logistic model derived from the Verhulst continuous logistic model. As its continuous cousin, the discrete logistic model only depends on one parameter. When this parameter is small, the system converges to a steady point. When it increases, there emerges a cycle of period two, which is new when you consider the continuous logistic equation. Thus continuous and discrete in time models behave differently. Let us go on and increase the parameter of the discrete logistic equation: we can observe the emergence of cycles with arbitrary lengths accompanied with sequences which densely fill an interval. Now there is no more connection with the continuous logistic model, no longer. Solutions to the discrete logistic model display astonishing features. In particular, some solutions are very sensitive to changes of the initial condition or to minor alterations of the parameter. Such models are called chaotic models. What about the biological implications of such mathematical results? Can we indeed model a real population with such a sensitive system? Would it be more realistic to resort to probabilistic modeling? Long-term predictions seem delicate or even impossible with such models. Can we accept models without predictions which constitute the essence of a scientific approach? Such issues are significant for the modeliser. Practically speaking, he has to choose between determinist and random models. Hence, we need an experimental approach which enables us to make such a discrete logistic model valid from a biological viewpoint. That is what [16] did through the study of the dynamics of *Tribolium*, which we will reproduce after the study of the discrete logistic model. [16] study a more sophisticated model but it presents the same bridges between steady states, cycles and chaotic behaviors. Such bridges will help make their model empirically valid and will enable us to stand in favor of modeling through a chaotic-type model.

3.2 Delay models

Let us start with the basic properties of difference equations. It is sometimes more natural, when modeling the evolution of a population, to take into account not only the current situation, but also the past one: for instance, birth rate does not depend on the population size N_n, but rather on the sexually mature individuals. A way to incorporate the delay effect is to consider models like:

$$N_{n+1} = f(N_n, N_{n-R}) , \qquad (3.1)$$

where R (an integer) is the delay. We will study the stability of the steady state of these delay models.

3.2.1 Case $R = 1$

Equation (3.1) becomes:

$$N_{n+1} = f(N_n, N_{n-1}) . \qquad (3.2)$$

Set $X_n = \begin{pmatrix} N_{n-1} \\ N_n \end{pmatrix}$. Equation (3.2) can be written:

$$X_{n+1} = F(X_n) , \qquad (3.3)$$

with

$$F\begin{pmatrix} x \\ y \end{pmatrix} = \begin{pmatrix} F_x(x, y) \\ F_y(x, y) \end{pmatrix} ,$$

$$F_x(x, y) = y \quad \text{and} \quad F_y(x, y) = f(y, x) .$$

A steady point of (3.3) is such that $X^* = \begin{pmatrix} x^* \\ y^* \end{pmatrix}$ with $x^* = y^*$ and $x^* = f(x^*, x^*)$. Indeed, this is a steady point of (3.2).
Let us set $\Sigma_n = X_n - X^*$ and linearize (3.3) about X^*:

$$\Sigma_{n+1} = M\Sigma_n + o(\Sigma_n) ,$$

with

$$M = \begin{pmatrix} 0 & 1 \\ \left(\frac{\partial f}{\partial y}\right)_{X^*} & \left(\frac{\partial f}{\partial x}\right)_{X^*} \end{pmatrix} .$$

The eigenvalues of M satisfies $P(\lambda) = 0$, where:

$$P(\lambda) = \lambda^2 - \lambda \left(\frac{\partial f}{\partial x}\right)_{X^*} - \left(\frac{\partial f}{\partial y}\right)_{X^*} .$$

The steady point X^* is stable if the modulus of the eigenvalues of M are less than 1.

3.2.2 General delay($R \geq 1$)

The approach is the same as for $R = 1$. A linearization about the steady point is done. The steady state is stable if the modulus of the roots of

$$P(\lambda) = \lambda^{R+1} - \lambda^R \left(\frac{\partial f}{\partial x}\right)_{X^*} - \left(\frac{\partial f}{\partial y}\right)_{X^*} \tag{3.4}$$

are all less than 1. A sufficient condition of stability is given by:

$$\left|\left(\frac{\partial f}{\partial x}\right)_{X^*}\right| + \left|\left(\frac{\partial f}{\partial y}\right)_{X^*}\right| < 1 .$$

Indeed, under this condition, the polynomial $P(\lambda)$ satisfies, for $|\lambda| \geq 1$:

$$|P(\lambda)| \geq |\lambda|^{R+1} - |\lambda|^R \left|\left(\frac{\partial f}{\partial x}\right)_{X^*}\right| - \left|\left(\frac{\partial f}{\partial y}\right)_{X^*}\right|$$

$$\geq |\lambda|^R \left(|\lambda| - \left|\left(\frac{\partial f}{\partial x}\right)_{X^*}\right| - \left|\left(\frac{\partial f}{\partial y}\right)_{X^*}\right|\right)$$

$$\geq 1 - \left|\left(\frac{\partial f}{\partial x}\right)_{X^*}\right| - \left|\left(\frac{\partial f}{\partial y}\right)_{X^*}\right| ,$$

and the modulus of the roots of $P(\lambda)$ are less than 1. Note that the value R of the delay disappears when using this (sufficient) condition.

3.2.3 Comparison with the system without delay

Let us consider the system without delay associated with the delay model (3.1):

$$N_{n+1} = f(N_n, N_n) . \tag{3.5}$$

The steady states of (3.1) and (3.5) are the same. A steady state of the system without delay is stable if:

$$\left|\left(\frac{\partial f}{\partial x}\right)_{X^*} + \left(\frac{\partial f}{\partial y}\right)_{X^*}\right| < 1 . \tag{3.6}$$

This condition is usually different from the condition given on the modulus of the roots of the polynomial (3.4). It is unclear whether the introduction of a delay stabilizes or destabilizes the models. Examples are given in exercise (3.6.1).

3.3 Discrete logistic model

Starting from a continuous-time model:

$$\frac{dN(t)}{dt} = f(N(t)) \,,$$

we introduce the "associated" discrete-time model:

$$\frac{N((n+1)\Delta) - N(n\Delta)}{\Delta} = f(N(n\Delta)) \,.$$

Most of the continuous-time models are based on the logistic function $f(x) = \rho x \left(1 - \frac{x}{\kappa}\right)$. The discrete-time logistic model is:

$$\frac{N((n+1)\Delta) - N(n\Delta)}{\Delta} = \rho N(n\Delta) \left(1 - \frac{N(n\Delta)}{\kappa}\right) \,.$$

Let us make the changes of parameters: $r = \rho\Delta + 1$, $K = \dfrac{\kappa(\rho\Delta + 1)}{\rho\Delta}$ and set $u_n = \dfrac{N(n\Delta)}{K}$. We then obtain:

$$u_{n+1} = r u_n (1 - u_n) \,. \tag{3.7}$$

This model have been introduced by [59] and [60] for modeling a population dynamics. From now on, we set:

$$\ell_r(x) = r x (1 - x) \,. \tag{3.8}$$

We need to work with a positive population size: we will thus assume in the following $0 \leq r \leq 4$ and $0 < u_0 < 1$. We can easily check that the population size remains in the interval $[0, 1]$.

3.3.1 Steady states

The sequence (3.7) has two steady states: 0 and $\dfrac{r-1}{r}$.

- $0 \leq r < 1$. The sequence u_n converges to 0.
- $1 < r < 3$. The state 0 becomes unstable. The state $\dfrac{r-1}{r}$ is stable. We check that the sequence u_n converges to $\dfrac{r-1}{r}$ for any initial condition $0 < u_0 < 1$.
- $3 < r \leq 4$. The steady states 0 and $\dfrac{r-1}{r}$ are both unstable.

  We will now study the case $3 < r \leq 4$.

3.3.2 Cycles

Consider the model where the iterative time step is 2:

$$u_{n+2} = \ell_r(\ell_r(u_n))$$
$$\equiv \ell_r^2(u_n) \, .$$

The steady states of the sequence u_{n+2}, apart from 0 and $\dfrac{r-1}{r}$, are:

$$u_{\pm} = \frac{r + 1 \pm \sqrt{(r+1)(r-3)}}{2r} \, . \tag{3.9}$$

This shows the existence of a discrete cycle of period 2: if $u_0 = u_+$, then $u_{2n} = u_+$ and $u_{2n+1} = u_-$. This is a first difference between continuous-time and discrete-time models. A one-dimensional continuous-time model has no periodic behavior; a discrete-time model can have a periodic behavior.

Definition 3.3.1 *Cycles.*
   Consider the iterative sequence $u_{n+1} = f(u_n)$. A cycle of period m is a sequence $c_0, c_1, \ldots, c_{m-1}$ such that:

$$c_i = f(c_{i-1}) \, ,$$
$$f^m(c_0) = c_0 \, ,$$
$$f^i(c_0) \neq c_0 \text{ for } i = 1, 2, \ldots, m-1 \, .$$

Proposition 3.3.1 *Stability of a cycle.*
A cycle is stable if $\left| \displaystyle\prod_{i=0}^{m-1} f'(c_i) \right| < 1 \, .$

Indeed, we know that the stability of the sequence $u_{n+m} = f^m(u_n)$ about c_i is given by the condition $|(f^m)'(c_i)| < 1$. But:

$$(f^m)'(c_i) = f'(f^{m-1}(c_i))(f^{m-1})'(c_i)$$
$$= f'(c_{i-1})(f^{m-1})'(c_i)$$
$$= \prod_{i=0}^{m-1} f'(c_i) \, .$$

We then check that the cycle (3.9) of period 2 of the discrete logistic map is stable if $3 < r < 1 + \sqrt{6}$ and unstable if $1 + \sqrt{6} < r \leq 4$. We then prove the existence of an increasing sequence r_n, with $r_n > 3$ and $\lim r_n = r_c \sim 3,828$, such that the associated discrete logistic map has cycles of period 2^n. To every r_n, a small interval is associated, for which the cycle of period 2^n is stable. We can prove that the sequence r_n satisfies:

$$\lim_{n \to \infty} \frac{r_n - r_{n-1}}{r_{n+1} - r_n} = \delta \sim 4.6692 \ldots$$

This constant δ is indeed an universal one ([22]). When $r > r_c$, the cycles of period 2^n become unstable and cycles of period $k, 2k, 4k, \ldots$, with k odd, appear. Note that a general result due to Sarkovsky ([77]) ensures that the existence of a cycle of period 3 implies the existence of cycles of period k, with k being an arbitrary integer. The existence of a cycle of period 3 therefore plays a key-role, for the existence of very disturbed behaviors, called chaotic behaviors ([55]). We will study these behaviors in the particular case $r = 4$.

3.3.3 Chaotic behavior

In this book, we do not want to give a survey of the general ergodic theory. We will only study the particular case:

$$u_{n+1} = 4u_n(1 - u_n) \,.$$

See [15] for general results, especially on the Birkhoff's Ergodic Theorem.

Let us write the initial condition: $u_0 = \sin^2(2\pi\theta)$. We can check that $u_n = \sin^2(2^n 2\pi\theta)$.

Let us decompose the real number θ in base 2:

$$\theta = \sum_{k \geq 0} \varepsilon_k 2^{-k} \,, \tag{3.10}$$

the ε_k being 1 or 0. The sequence u_n becomes:

$$u_n = \sin^2 \left(2\pi \sum_{p > 0} 2^{-p} \varepsilon_{p+n} \right) \,. \tag{3.11}$$

When θ is a rational number, the expansion (3.10) is periodic after some rank. Equation (3.11) shows that the sequence u_n is periodic after this rank. Cycles of arbitrary periods can be built from a rational initial condition.

Two remarks can be done when starting from a rational number θ.

- *Sensitivity to the initial condition.* Let θ_1 and θ_2 be two rational numbers such that $|\theta_1 - \theta_2| \leq 2^{-K}$ with K huge, but with binary expansions that differ after the K-th digit. The corresponding logistic maps u_n^1 and u_n^2 will be closed at the beginning (roughly speaking for $n < K$), but will then have completely different behaviors.
- *Computer simulation.* Let us simulate the logistic map with a computer. The initial condition θ has to be a decimal number in base 2. The simulated sequence u_n will therefore equal to 0 after some rank.

We will show that there exists a set K, which complement is negligible, such that, if $\theta \in K$, then the sequence u_n visits the interval $[0, 1]$ in a dense way. Moreover, we will give the density occupation of the sequence u_n.

We will first check that the sequence of functions:

$$V_n(\theta) = \frac{1}{n} \sum_{k=0}^{n-1} \exp(i2\pi\theta 2^n k)$$

with k being an non-vanishing integer, converges in L^2 to the null function since $\int_0^1 |V_n(\theta)|^2 d\theta = \frac{1}{n}$. Then little algebra proves that there exists a constant C such that $\int_0^1 |V_n(\theta)|^4 d\theta \leq \frac{C}{n^2}$. Set $A_n = \{\theta \in [0,1], |V_n(\theta)| \geq \varepsilon\}$. Denote by $\lambda(A_n)$ the Lebesgue measure of A_n.

$$\lambda(A_n) = \int_0^1 \mathbf{1}_{|V_n(\theta)|\geq\varepsilon} d\theta$$

$$= \frac{1}{\varepsilon^4} \int_0^1 \varepsilon^4 \mathbf{1}_{|V_n(\theta)|\geq\varepsilon} d\theta$$

$$\leq \frac{1}{\varepsilon^4} \int_0^1 |V_n(\theta)|^4 d\theta$$

$$\leq \frac{C}{\varepsilon^4 n^2} .$$

Set $B_n = \cup_{p\geq n} A_p$. We then have $\lim_{n\to+\infty} \lambda(B_n) = 0$. There follows that $V_n(\theta) \to 0$ (a.e.) as $n \to +\infty$. Writing $\sin(x) = \dfrac{e^{ix} - e^{-ix}}{2i}$, we can then deduce

$$\frac{1}{n} \sum_{k=0}^{n-1} u_k^p \to \frac{C_{2p}^p}{2^{2p}} \text{ (a.e.) ,}$$

as $n \to \infty$, and for p being an arbitrary integer. Particularly, for $p = 1$, it means that there exists an average size. This average size is equal to $1/2$. Set:

$$\mu(x) \equiv \frac{1}{\pi\sqrt{x(1-x)}} .$$

Noting that $\int_0^1 x^p \mu(x)dx = \dfrac{C_{2p}^p}{2^{2p}}$, Stone-Weierstrass Theorem ensures that, for every continuous function, we have:

$$\frac{1}{n} \sum_{k=0}^{n-1} f(u_k) \to \int_0^1 f(x)\mu(x)dx \text{ (a.e.) ,} \tag{3.12}$$

as $n \to \infty$. This convergence (3.12) is still available for indicator function $\mathbf{1}_{[a,b]}$, $0 \leq a < b \leq 1$:

$$\frac{1}{n} \sum_{k=0}^{n-1} \mathbf{1}_{[a,b]}(u_k) \to \int_a^b \mu(x)dx \text{ (a.e.) .} \tag{3.13}$$

The convergence (3.13) clearly indicates that μ is the occupation density of the sequence u_n. Moreover, the convergence (3.12) ensures that there is no stable cycle. Indeed, assume the existence of a stable cycle: $c_0 = \sin^2(2\pi\theta_0), c_1 = \sin^2(2\pi\theta_1), \ldots, c_{m-1} = \sin^2(2\pi\theta_{m-1})$. Then there exists a small interval I (of non-vanishing Lebesgue measure) about θ_0 such that, for any initial condition $\theta \in I$, we have:

$$\frac{1}{n} \sum_{k=0}^{n-1} f(u_k) \rightarrow \frac{1}{m} \sum_{i=0}^{m-1} f(c_i) \text{ as } n \rightarrow \infty.$$

This clearly is a contradiction with (3.13).

Moreover, note that the density μ is infinite in 0 and 1. This indicates that the population can almost vanish from time to time, but that this population will nevertheless never completely vanish.

Remark 3.3.1 *Comparison with a purely stochastic model.*

Assume that the population size is driven by a purely stochastic model, say the sequence u_n is an i.i.d. sequence of r.v. of probability density μ. The strong law of large number shows that, for every function f:

$$\frac{1}{n} \sum_{k=0}^{n-1} f(u_k) \rightarrow \int_0^1 f(x)\mu(x)dx \text{ (a.s.)} .$$

This result is similar to (3.12).

The three following figures enables us to graphically understand the apparition of cycles for the discrete logistic model.

3.4 *Tribolium* **dynamics.**

We have seen that an iterative discrete map, even very basic, as in the logistic map, can lead to very complicated behaviors, especially chaotic behaviors. The following question then arises. Is there such (mathematical) behavior in the (biological) real world? The distinction between a chaotic determinist behavior and a stochastic behavior seems particularly delicate to establish empirically. Indeed, a real situation always contains uncertainties: some parameters are unknown, some data are noisy, ... When the model has a stable behavior (*e.g.* stable steady state or stable cycles), such uncertainties do not play a major role. On the other hand, when the behavior is unstable (the sensitivity to initial conditions is then essential), such uncertainties play a key-role.

[16] have elaborated an experimental protocol. This protocol is a way of deciding whether a population dynamics is determinist (chaotic) or stochastic. [16] have studied the dynamics of the coleopter *Tribolium*. There are three stages in the life of a *Tribolium*: worm, chrysalis and adulthood. Not taking

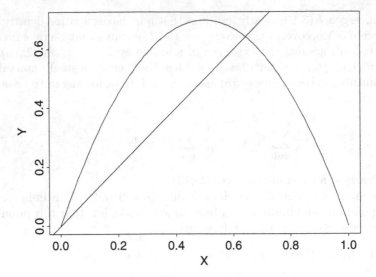

Fig. 3.1. Function $\ell_r(x)$ for $r = 2.8$.

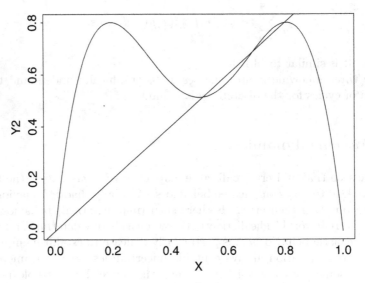

Fig. 3.2. Function $\ell_r^2(x)$ for $r = 3.2$.

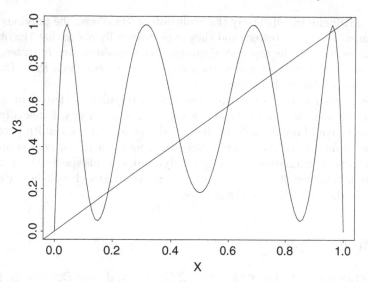

Fig. 3.3. Function $\ell_r^3(x)$ for $r = 3.95$.

into account the errors of measures, the stochastic variability of the various rates (births, deaths, ...) and the unexpected changes in the outside world, the proposed model is the following:

$$W_{n+1} = rA_n \exp(-c_{e,w}W_n - c_{e,a}A_n), \tag{3.14}$$
$$P_{n+1} = (1 - \mu_w)W_n,$$
$$A_{n+1} = P_n \exp(-c_{c,a}A_n) + (1 - \mu_a)A_n.$$

W_n represents the worms that eat; P_n represents the great worms, the worms that do not eat, the chrysalis and the sexual immature adults; A_n represents the sexually mature adults. The time unit is two weeks. This is approximatively the time a *Tribolium* spends in stages W_n and P_n. The rate r is the reproduction rates per adult by time unit, and without cannibalism. The rates μ_w and μ_a are the death rates of worms and adults by time unit. The parameter $\exp(-c_{e,w}W_n - c_{e,a}A_n)$ models the eggs eaten by worms and adults. The parameter $\exp(-c_{c,a}A_n)$ models the chrysalis eaten by the adults.

When the parameters $r, -c_{e,w}, c_{e,a}, \mu_w$ and μ_a are given (only parameter $c_{c,a}$ is allowed to vary), the model (3.14) has the following behavior:

- $c_{c,a} = 0.0$. One stable steady state.
- $c_{c,a} = 0.05$. A stable cycle of period 8.
- $c_{c,a} = 0.25$ and $c_{c,a} = 0.35$. Chaotic behavior.
- $c_{c,a} = 0.423 \to 0.677$. Cycle of period 3 together with chaotic behavior and stable cycles of period 8 and more.
- $c_{c,a} = 1.0$. Stable cycle of period 3.

Then [16] empirically study the evolution of *Tribolium*: the previous values of parameters $c_{c,a}$ are tested, and they experimentally check that the observations correspond to the theoretical model. The transition steady states, cycles and chaos is observed. It seems then reasonable to conclude that *Tribolium* should have a chaotic behavior.

This experiment is a confirmation that a population driven by a small number of parameters can behave in a very complicated way. Especially, these fluctuations could not come from the outside world, but are resulting from the intrinsic dynamics of the species. Moreover, an even minor change in these parameters can dramatically modify the dynamics of this species. Last, despite a common belief (*cf.* [60]), this experiment proves that a chaotic behavior does not imply the extinction of the species.

3.5 Bibliography

* COSTANTINO, R., DESHARNAIS, R., CUSHING, J. and DENNIS, B. (1997). Chaotic dynamics in an insect population. *Science.* **275** 389-391.
* FEIGENBAUM, M. (1978). Quantitative universality for a class of nonlinear transformations. *J. Stat. Phys.* **19** 25-52.
* LI, T. and YORKE, J. (1975). Period three implies chaos. *Amer. Math. Monthly.* **82** 985-992.
* MAY, R. (1976). Simple mathematical models with very complicated dynamics. *Nature.* **261** 459-467.
* MAY, R. and OSTER, G. (1976). Bifurcations and dynamic complexity in simple ecological models. *Am. Nat.* **110** 573-599.
* MURRAY, J. (1990). *Mathematical Biology.* Springer-Verlag.
* SARKOVSKY, A. (1964). Coexistence of cycles of a continuous map of a line into itself (in Russian). *Ukr. Mat. Z.* **16** 61-71.

3.6 Exercises

Exercise 3.6.1 *Delay models.*

Compare the stability of the steady states in the following models:

$$u_{n+1} = f(u_n, u_{n-1}),$$
$$v_{n+1} = f(v_n, v_n),$$

with $f(x, y) = x \exp(r(1 - y))$, and then $f(x, y) = \dfrac{rx^2}{1 + by^2}$. Parameters r and b are positive.

Exercise 3.6.2 *Cycles of period 2.*

For which values of parameter r are there cycles of order 2? Are the cycles stable?

$$u_{n+1} = u_n \exp(r(1 - u_n)) .$$

Exercise 3.6.3 *Two species model.*

Determine the stability of the (possible) steady state in the model:

$$N_{n+1} = f(N_n, P_n) ,$$
$$P_{n+1} = g(N_n, P_n) ,$$

where f and g are positive functions.

Application: Predator-prey model with $f(x, y) = rx \exp(-ay)$ and $g(x, y) = \rho x(1 - \exp(\alpha y))$ where a, α, r and ρ are positive.

Exercise 3.6.4 *Dispersal model (from [50]).*

Let $N_0(x), x \in \mathbb{R}$ be a population density: the function $N_0(x)$ is positive and its integral is equal to N_0. Let k be a positive kernel which integral is equal to 1. Assume the following repartition of the population through time:

$$N_{n+1}(x) = \rho \int_{\mathbb{R}} k(x - y) N_n(y) dy .$$

1. Justify this dispersal model.
2. Calculate the total population size at time n.
3. Can this model generate travelling waves with an accurate choice of parameter ρ and kernel k,? In other words, can we find a constant c such that:

$$N_{n+1}(x) = N_n(x - c) ?$$

4. Now we assume that the kernel k and the function N_0 are compactly supported. What is the colonization speed of $\mathbb{R}$ by the population?
5. Now assume a periodic model::

$$N_{n+1}(x) = \int_0^{2\pi} k(x - y) N_n(y) dy .$$

Justify this model. What is the limit of $N_n(x)$ as $n \to +\infty$?

Exercise 3.6.5 *Optimal harvesting (e.g. [12, Ch.18.3]).*

Consider a population $u_n, n \geq 0$ to be harvested. Its life cycle is described by a set of stages (*i.e.* i-states, $i = 1, \ldots, d$): $u_n = (u_n^i)_{i=1,\ldots,d}$. We assume the transitions between the different stages to be linear:

$$u_{n+1} = Au_n \,,$$

where $A = (a_{ij})_{i,j=1,\ldots,d}$.

Let h_i be the proportion of stage i surviving the harvest. The model becomes:

$$u_{n+1} = HAu_n \,,$$

with $H = diag(h_1, \ldots, h_d)$.

1. Let y be the yield of an individual in each stage. The total yield is:

$$Y = y^t(I - H)Au \,.$$

We want to maximize the yield per individual. Show that this maximization reduces to the linear programming problem:
Find u that maximizes

$$Y = y^t(A - I)u \,,$$

subject to the constraints

$$Au - u \geq 0 \,,$$
$$u \geq 0 \,,$$
$$(1, \ldots, 1) \cdot u = 1 \,.$$

2. Find H in terms of the optimal vector u.

4

Game theory and evolution

4.1 Introduction

This section does not pretend to propose a thorough introduction to game theory but only to give an idea of the application of simple games to the theory of the evolution. From a historical viewpoint, the fields of application of the game theory is economy, and it is so since the fundamental writings of [85]. A game is a mathematical object which presents a conflict between several subjects called players. It is up to the players to choose their strategy in the game. From a mathematical viewpoint, the players try to make the most of the numerous strategies. Hence the problem of an equilibrium in the game combining each strategy of each player. The game theory needs a significant theoretical investment in non-linear analysis, which we will not make. You can read general books such as [3, 30, 71, 81, 87]). We will only set the definitions of essential notions such as game, strategy and equilibrium and we will only speak about, and not demonstrate, the Nash Theorem which proves the existence of at least one equilibrium.

The application of the game theory to the evolution is much more recent than its application to economy ([61]). From the point of view of evolution, two issues are commonly dealt with: the sex-ratio and the hawk-dove model. [23] write about the stakes of both issues from a neo-Darwinian viewpoint, in the light of the game theory. We chose to thoroughly deal with the hawk-dove model. The problem is as follows: when observing the behavior of some animal species in case of internal conflict (*i.e.* the appropriation of a new food resource or the conquest of a female), there are at least two behaviors. The first one is fight: the aim is to obtain this resource at any cost, no matter what the consequences are. Such behaviors is called the "hawk" behavior. The second one is peace: you 'd rather lose the resource than risk a conflict. This second behavior is called the "dove" behavior. The conflict between the two protagonists will be modeled in terms of games: the players will stand for the animals and the strategies will be the hawk and dove behaviors. The questions is: how will the parameters of the game (significance of the resource and risks of corporal damages) make a balance possible between hawks and doves within a population? Firstly, we will study the static aspect of the game. The dynamical aspect is still in progress from a mathematical viewpoint. We will propose a dynamical version of this game, which will allow an interesting parallel with the dynamic demographical models. Finally we will propose exercises as an extension of the game: the hawk-dove-bourgeois game.

4.2 Games, strategies and equilibria

We will first start with a formal presentation of the notions of games, strategies and equilibria. Indeed, we will only consider non-cooperative games, and the word "non-cooperative" will be omitted for simplicity. We consider two players A and B. These players would model either individuals, either populations. These two players are in competition. The aim of player A (resp. B) is to chose a strategy x (resp. y) in a set of strategies. For convenience, we denote by A (resp. B) the set of strategies of player A (resp. B). In the sequel we will give examples of strategies: the celebrated hawk versus dove strategy and in exercise 4.5.1 one of its generalization, the hawk-dove-bourgeois game. The set $A \times B$ is the set of possible strategies. By assumption, both sets A and B are finite, but infinite sets of strategies can be considered. The elements (=strategies) of A and B are denoted by A_i and B_j. The players can therefore chose between a finite number of strategies. The players A and B do not know about the opponent's choice.

Let us follow the classical approach. We assume that the players do not chose their strategies, but only chose the probability of playing a strategy. Therefore let us define the subset $\overline{A}$ of $\mathbb{R}^{\#A}$:

$$\overline{A} = \{x \in (\mathbb{R}^+)^{\#A}, \sum_1^{\#A} x_i = 1\} .$$

The set A is then "embedded" in the set $\overline{A}$: the $i-th$ element of A is replaced by the vector $(0, \ldots, 1, \ldots, 0)$, with the 1 in $i-th$ position.

Therefore we have obtained a continuous set $\overline{A}$ of strategies. These strategies are called *mixed* strategies as opposed to the strategies of A that are called *pure* strategies. A mixed strategy is a convex combination of pure strategies: each weight of the convex combination can be interpreted as the probability that the player A chose the pure strategy associated with the weight. Of course, the same holds for B, and a set $\overline{B}$ of mixed strategies is defined for the player B.

A game is then defined by a function f from $A \times B$ to $\mathbb{R}^2$, called a utility function. The function f associates each pair $(x, y) \in A \times B$ of (pure) strategies with a pair $(f_A(x, y), f_B(x, y)) \in \mathbb{R}^2$. $f_A(x, y)$ represents the "profit" of player A if he plays the strategy x with his opponent B playing the strategy y.

Function f is then defined on the sets $\overline{A}$ and $\overline{B}$ by bi-linearization. Let $x = (x_i)$ and $y = (y_j)$ be mixed strategies of $\overline{A} \times \overline{B}$.

Then:

$$f_A(x, y) = \sum_{i,j=1}^{\#A, \#B} x_i y_j f_A(A_i, B_j) ,$$

$$f_B(x, y) = \sum_{i,j=1}^{\#A, \#B} x_i y_j f_B(A_i, B_j) .$$

Of course the behaviors of players A and B result from the utility function. Player A choses strategy $x^\star$ that maximizes his profit, but he still ignores the strategy of player B. In other words, player A choses a strategy $x^\star$ in the set

$$\{x^\star \in \overline{A}, \ f_A(x^\star, y) = \max_{x \in \overline{A}} f_A(x, y)\} .$$

Similarly, player B choses a strategy $y^\star$ in the set

$$\{y^\star \in \overline{B}, \ f_B(x, y^\star) = \max_{y \in \overline{B}} f_B(x, y)\} .$$

Player B also ignores the strategy of A. The question is now to know whether there exists a pair of strategies $(x^\star, y^\star)$ satisfying the previous conditions. Such a pair is called a Nash equilibrium of the game. A pair of strategies $(x^\star, y^\star)$ is a Nash equilibrium of the game iff:

$$f_A(x^\star, y^\star) = \max_{x \in \overline{A}} f_A(x, y^\star) ,$$

$$f_B(x^\star, y^\star) = \max_{y \in \overline{B}} f_B(x^\star, y) .$$

The question of the existence of a Nash equilibrium is a difficult question. It relies on the Brouwer fixed point Theorem. We will therefore admit the results without any proof. Indeed, we will give results on Nash equilibrium in a more general setup than those used in the examples. These results can be particularly used when the number of strategies becomes infinite.

The sets $\overline{A}$ and $\overline{B}$ are convex and compact. The functions f_A and f_B are continuous. We are in a particular case of the general Nash Theorem ([87],[3, Ch.12]).

Theorem 4.2.1 *Nash Theorem.*

Assume that the functions $x \to f_A(x, y)$ and $y \to f_B(x, y)$ are concave[1], then the game has at least one Nash equilibrium.

---

[1] In our examples, this condition is still satisfied.

4.3 Hawks and doves

4.3.1 Equilibria

We will consider an animal population with infinite size. These animals are in competition for a resource, for instance food. Inside this population, two behaviors coexist: a behavior H (for hawks) and a behavior D (for doves). The meetings between hawks and doves are characterized as follows.

1. When two doves meet, they fairly share the resource R.
2. When a hawk and a dove meet, the hawk brings the resource R without fight.
3. When two hawks meet, they fight together. This induces damages P and each hawk bring only half of the resource R minus the damages P.

We assume that the resource R and the damages P are of the same kind. These two quantities are of different kind through this means that they have been converted in terms of *selective value*. Let $G(.,.)$ be the "profit" of a meeting. Depending of the kind of meeting, this profit is given by:

$$G(H,H) = \frac{R-P}{2} ,$$
$$G(H,D) = R ,$$
$$G(D,H) = 0 ,$$
$$G(D,D) = \frac{R}{2} .$$

This is a game with pure strategies: the hawk strategy and the dove strategy. Let λ be the proportion of doves in the population. This game can be viewed as a two-player-game: λ is the probability for a player to adopt the dove strategy. With the previous notation, the utility function $f = (f_A, f_B)$ becomes:

$$f_A(H,H) = f_B(H,H)$$
$$= G(H,H) ,$$
$$f_A(H,D) = f_B(D,H)$$
$$= G(H,D) ,$$
$$f_A(D,H) = f_B(H,D)$$
$$= G(D,H) ,$$
$$f_A(D,D) = f_B(D,D)$$
$$= G(D,D) .$$

Let us now investigate the Nash equilibrium of this game. Let us compute $\max_{x \in \overline{A}} f_A(x,y)$. The strategy x is a mixed strategy and can be written $x = \lambda C + (1-\lambda)F$. We then deduce:

$$\max_{x \in \overline{A}} f_A(x, y) = \max(f_A(D, y), f_A(H, y)) .$$

The strategy y is also a mixed strategy: $y = \lambda' D + (1 - \lambda')H$. So:

$$\max_{x \in \overline{A}} f_A(x, y) = \max\left(\lambda'\frac{R}{2}, \lambda' R + (1 - \lambda')\frac{R - P}{2}\right) .$$

Two cases must be distinguished: when the damage P is greater than the resource R, and when the damage P is less than the resource R.

- Case $R > P$.
  We have $f_A(H, y) > f_A(D, y)$ and:

$$\max_{x \in \overline{A}} f_A(x, y) = f_A(H, y) .$$

Because of the symmetry between the functions f_A and f_B, the equilibrium $(x^\star, y^\star)$ satisfies $x^\star = y^\star$. We are looking for an $x^\star$ such that:

$$f_A(x^\star, x^\star) = f_A(H, x^\star) .$$

Set $x^\star = \lambda^\star D + (1 - \lambda^\star)H$.

$$f_A(x^\star, x^\star) = f_A(H, x^\star)$$
$$= \lambda^\star f_A(D, x^\star) + (1 - \lambda^\star)f_A(H, x^\star) .$$

We can deduce $\lambda^\star = 0$. The strategy hawk-hawk is the Nash equilibrium of the game.

- Case $R < P$.
  We have just seen that if $\max_{x \in \overline{A}} f_A(x, y)$ equals $f_A(H, y)$, then $\lambda^\star = 0$. It follows $\max_{x \in \overline{A}} f_A(x, y) = \max\left(\dfrac{R - P}{2}, 0\right)$. We can deduce that $\max_{x \in \overline{A}} f_A(x, y)$ equals $f_A(D, y)$ in the case $R < P$. We have to solve:

$$f_A(D, x^\star) = f_A(H, x^\star) . \tag{4.1}$$

The frequency $\lambda^\star$ defined by the equation (4.1) is a frequency that makes the balance between the profits of hawks and doves. This frequency is given by:

$$\lambda^\star = 1 - \frac{R}{P} .$$

The Nash equilibrium leads to a proportion $1 - \dfrac{R}{P}$ of doves and a proportion $\dfrac{R}{P}$ of hawks. The game theory modeling explains the coexistence of the two behaviors when the hawk's strategy is too costly.

4.3.2 Dynamical aspect

When a game has an equilibrium, the question of its stability arises. For instance, what happens if some immigrants or mutants arrive? Let us study the stability of the hawk-dove game. By stability, we mean stability in time. For this purpose, we need a reproduction model of the population. We can assume, even if it is unrealistic, the most simple reproduction model: the asexual reproduction model. Moreover, we want an expected number of descendants per individual, that is consistent with the hawk-dove game.

Let λ_n be the frequency of doves in the population at time n. A dove has an expected profit given by:

$$G_n(D) = \lambda_n G(D, D) + (1 - \lambda_n)G(D, H) .$$

A hawk has an expected profit given by:

$$G_n(H) = \lambda_n G(H, D) + (1 - \lambda_n)G(H, H) .$$

We assume that the reproduction is proportional to the profit:

$$\lambda_{n+1} = \lambda_n \frac{G_n(D)}{\lambda_n G_n(D) + (1 - \lambda_n)G_n(H)} . \tag{4.2}$$

The steady states of the sequence λ_n are the solutions of:

$$\lambda^\star = \lambda^\star \frac{G^\star(D)}{\lambda^\star G^\star(D) + (1 - \lambda^\star)G^\star(H)} .$$

The two solutions are $\lambda^\star = 0$ and $\lambda^\star = 1 - \dfrac{R}{P}$. The second solution is admissible iff $R < P$. The steady states of the sequence λ_n are the Nash equilibrium of the game. The equation (4.2) can be written:

$$\lambda_{n+1} = h(\lambda_n) ,$$

where

$$h(x) = \frac{Rx^2}{Rx^2 + 2Rx(1 - x) + (R - P)(1 - x)^2} .$$

Let us study the stability of the steady states of the sequence λ_n.

1. Neighborhood of $\lambda^\star = 0$.
   A Taylor expansion leads to:

$$\lambda_{n+1} \sim \lambda_n^2 \frac{R}{R - P} .$$

   a) If $R > P$, $\lambda_n \to 0^+$ as $n \to +\infty$ and the state $\lambda^\star = 0$ is stable.
   b) If $R < P$, $\lambda_n \to 0^-$ as $n \to +\infty$ and the model is not admissible for populations having a small number of doves.

2. Neighborhood of $\lambda^\star = 1 - \dfrac{R}{P}$.

We check that $h'(\lambda^\star) = 2\lambda^\star - 1$. The steady state $\lambda^\star$ is stable.

When $R > P$, the state $\lambda^\star = 0$ is stable, one says that the strategy hawk-hawk is an ESS, where ESS stands for Evolutionary Stable Strategy. When $R < P$, the state $\lambda^\star = 1 - \dfrac{R}{P}$ is stable, one says that the mixed strategy is an ESS. These strategies are the unique ESS's of the game.

4.4 Bibliography

- AUBIN, J.-P. (1998). *Optima and Equilibria: An Introduction to Nonlinear Analysis*. Springer.
- GOUYON, P.-H., HENRY, J.-P. and ARNOULD, J. (2002). *Gene Avatars. The Neo-Darwinian Theory of Evolution*. Kluwer.
- MAYNARD-SMITH, J. (1982). *Evolution and the Theory of Games*. Cambridge University Press.
- OWEN, G. (1982). *Game Theory*. Academic Press, 2nd edition.
- NEUMANN, J., MORGENSTERN, O. (1944). *Theory of Games and Economic Behavior* . Princeton : Princeton University Press.
- WEIBULL,J. (1995). *Evolutionary Game Theory*. MIT Press.

4.5 Exercises

Exercise 4.5.1 *Hawks, doves and bourgeois.*

Let us consider the hawk-dove game with a third possible strategy B: the bourgeois strategy. A bourgeois behaves like a hawk if he think he was first, else he behaves like a dove. The meeting of a bourgeois with an individual are modeled as follows:

1. When two bourgeois meet, they do not fight. One of them (the first arrived) takes the resource.
2. When a bourgeois meet a dove, they do not fight. If the bourgeois was first, he takes the resource. If not (the dove was first), the dove takes the resource.
3. When a bourgeois meets a hawk, he behaves like a hawk if he was first, else he behaves like a dove.

1. Give the utility function of the game.
2. Study the Nash equilibrium in terms of the resource R and damages P.
3. Propose a reproduction model associated with the game.

Exercise 4.5.2 *Absent-minded driver.*

Two cars arrive at a crossroads. If the two cars stop, the profit of the drivers is null. If the two drivers do not stop, the profit is negative (accident!) and huge. If one driver stops and the other doesn't, those who stops has a small negative profit and the other has a small positive profit. Model and study this game.

Exercise 4.5.3 *Prisoner's dilemma.*

Two suspects of a crime are arrested by the police. The police do not have enough evidence to convict either of them unless one of the suspects confesses. The police detain the suspects in individual cells and explain the consequences of their potential actions. If neither prisoner confesses, both are sentenced to one year in prison. If both confess, they will be imprisoned for six years. If only one of the confesses, then that prisoner will be released immediately while the other will be sentenced to nine years in prison. Model and study the game.

5

Markov chains and diffusions

5.1 Introduction

When faced with a concrete situation, the modeliser will have to ask himself this very important question: better a determinist model or a random one? So far we favored the determinist one, but there was no ideological bias. From now on, this book will be devoted to random model. We do not aim at questioning the nature of hazard in life sciences but we intend to account for phenomena which are better answered for thanks to a stochastic model than a determinist one.

This section will be devoted to dynamical random models and, particularly, to one of their sub-classes, the Markov[1] chains. The Markov chains model dynamical random phenomena in which the past only intervenes in the last moment of the chain. It is called random phenomena with short memory. When the Markov chain occurs in a finite or countable cardinal state-space, the study of the chain does not require any important mathematical investment, that is why we will start with such a case.

However, the study of Markov chains is undoubtedly tiresome to a certain extent, but we cannot avoid it. As soon as we have mentioned the definitions and first properties of the Markov chains, we will study the issue of genetical drift. In 1908, G. Hardy, a British mathematician, and W. Weinberg, a German physician, established that an isolated population living under reasonable conditions (that is, without mutation or selection due to an external factor, or immigration) has a constant genetical composition in time. Their basic proof lies on a determinist model of reproduction with two alleles. We considered the same biological assumptions, minus the fact that reproduction is now considered as random. So, we can model our isolated population with a Markov chain in which one of the two alleles always ends up vanishing, hence the term genetical drift that was applied to this model.

---

[1] Andrei Markov, 1856-1922, Russian probabilist.

Both models -determinist and stochastic- lead to contradictory conclusions. The stochastic model seems closer to the biological reality in this case and genetical drift is undoubtedly a seminal case which justifies the introduction of stochastic model.

Dynamical determinist systems offer two aspects: one is continuous-time, the other discrete-time. Similarly the random dynamical systems accept a continuous-time and a discrete-time pattern. The study of Markov continuous-time chain is much more delicate than the study of discrete-time processes and we will only go through it. We will start the study of continuous-time Markov processes with a building of the Brownian motion based on a random walk. Of course, such random walks will be modeled with a discrete-time Markov chain and the Brownian motion will be obtained as the double-renormalizations of the random walk, both in time and space. Incidentally, let us recall that the origins of the Brownian motion were found in the observations R. Brown[2] did in 1827: he discovered the unorganized motion (which bears his name) of very small particles in suspension in fluids. It must be the oldest model of a biological phenomena via a continuous-time random process. Let us not get carried along too far, though, since it will take almost a century before the Brownian motion is built with method as a mathematical object. Such construction of the Brownian motion will help us tackling the diffusion processes. Diffusion processes -as stochastic differential equations, but we will not mention them- belong to a large field of probability theory, which we will not be analyzed in detail. We will just give the features of the diffusion processes and precise some basic properties belonging to these stochastic processes.

Lastly we will use a whole concrete example: that of the domestication of pearl millet, due to [52] (see [20] for a survey on domestication, and more!). Its goal is to understand how cereals were domesticated in spite of other wild or hybrid species. Such model will enable us to go through the two types of dynamical approaches we met and will make us once again wonder about the use of the introduction of the stochastic factor in a model. We will use the determinist model with a differential system and then a stochastic model with a Markov chain. Finally, we will roughly approximate the Markov chain with a diffusion process and see that some parameters in the Markov chain (here the invariant measure of the Markov chain), which seem impossible to obtain via a direct calculation, can be approximately known thanks to the underlying diffusion processes. To conclude, we will provide each sub-chapter with the bibliography necessary to study the subject more thoroughly.

5.2 Definitions and first properties

We first consider discrete-time and discrete-state Markov chains X_n, $n \in \mathbb{N}$: X_n has only a finite or countable number of possible values of a set $\mathcal{S}$. See [29] for basic results on Markov chains and [62] for more general results.

---

[2] Robert Brown, 1773-1858, Scottish botanist.

5.2.1 Markov property

Definition 5.2.1 *Markov property.*
A random sequence $X_n, n \geq 0$, with values in a finite or countable state space $\mathcal{S}$ satisfies the Markov property if:

$$\mathbf{P}(X_{n+1} = x_{n+1}|X_0 = x_0, \ldots, X_n = x_n) = \mathbf{P}(X_{n+1} = x_{n+1}|X_n = x_n) \, .$$

The sequence X_n is then called a Markov chain.

The transition probability associated with a Markov chain is the sequence of all the probabilities of moving from a state to another:

$$\mathbf{P}(X_{n+1} = x|X_n = y) \ \ x, y \in \mathcal{S} \, .$$

A Markov chain is *homogeneous* if its transition probabilities do not depend on time. We then denote:

$$\begin{aligned} P(y,x) &= \mathbf{P}(X_{n+1} = x|X_n = y) \\ &= \mathbf{P}(X_1 = x|X_0 = y) \, . \end{aligned}$$

We only study homogeneous Markov chains and the term "homogeneous" will be omitted.

The transition probability clearly satisfies:

$$P(x,y) \geq 0 \ \forall (x,y) \in \mathcal{S}^2 \, ,$$
$$\sum_{y \in \mathcal{S}} P(x,y) = 1 \ \forall x \in \mathcal{S} \, .$$

The initial distribution of the Markov chain is the distribution π_0 of X_0:

$$\pi_0(x) = \mathbf{P}(X_0 = x) \, .$$

The initial distribution and the transition probability entirely characterizes the Markov chain.

The joint distribution of $X_0, \ldots, X_n$ can be recursively computed by Bayes formula:

$$\begin{aligned} \mathbf{P}(X_0 = x_0, X_1 = x_1) &= \mathbf{P}(X_0 = x_0)\mathbf{P}(X_1 = x_1|X_0 = x_0) \\ &= \pi_0(x_0)P(x_0, x_1) \, , \end{aligned}$$

and

$$\mathbf{P}(X_0 = x_0, \ldots, X_n = x_n) = \pi_0(x_0)P(x_0, x_1)\ldots P(x_{n-1}, x_n) \, .$$

The joint distribution of $X_{n+1}, \ldots, X_{n+m}$ conditionally to X_n can be similarly computed:

$$\mathbf{P}(X_{n+1} = x_{n+1}, \ldots, X_{n+m} = x_{n+m} | X_n = x_n) = P(x_n, x_{n+1}) \cdots$$
$$\cdots P(x_{n+m-1}, x_{n+m}).$$

The probability $P^m(x, y)$ of moving from the state x to the state y in m steps is therefore given by:

$$P^m(x, y) = \sum_{y_1 \in S, \ldots, y_m \in S} P(x, y_1) P(y_1, y_2) \ldots P(y_m, y).$$

Therefore, the sequence of probability P^m satisfies the Chapman-Kolmogorov equations:

$$P^{n+m}(x, y) = \sum_{z \in S} P^n(x, z) P^m(z, y).$$

Especially:

$$\mathbf{P}(X_n = x) = \sum_{y \in S} \pi_0(y) P^n(y, x). \tag{5.1}$$

5.2.2 Return time

An essential definition for understanding Markov chain is the return time into a subset of the state space S.

Definition 5.2.2 *Return time.*
Let A be a subset of the state space S. The return time to A is the random variable defined by:

$$T_A = \inf\{n \geq 1, \ X_n \in A\}.$$

By convention, when the sequence X_n never return to A, we set $T_A = +\infty$. When the set A is a singleton set $\{x\}$, we set $T_A = T_x$.

Some states are "killing" the Markov chain, they are called absorbing states.

Definition 5.2.3 *Absorbing states.*
A state x of the state space S is absorbing if $P(x, x) = 1$.

5.2.3 Finite state space

When the state space S is finite, the study of Markov chains can be done in a matricial framework.

Definition 5.2.4 *Transition matrix.*
    The transition matrix is the matrix $P = (P(x, y))$, $(x, y) \in \mathcal{S}^2$.

Let $\#\mathcal{S} = s$. Denote by π_0 the vector $(\pi_0(1), \dots, \pi_0(s))$ and by π_n the vector $(\mathbf{P}(X_n = 1), \dots, \mathbf{P}(X_n = s))$, $n \geq 1$. The relationship (5.1) becomes:

$$\pi_n = \pi_0 P^n .$$

5.2.4 State classification

Let us come back to the general framework: the state space $\mathcal{S}$ is finite or countable. For $(x, y) \in \mathcal{S}^2$, denote:

$$\rho_{x,y} = \mathbf{P}(T_y < \infty | X_0 = x) .$$

$\rho_{x,y}$ represents the probability that the Markov chain starting from the state x reaches the state y in a finite time.

Definition 5.2.5 *Recurrent and transient states.*
    A state y is recurrent if $\rho_{y,y} = 1$. A state y is transient if $0 \leq \rho_{y,y} < 1$.

If the state y is absorbing, $\mathbf{P}(T_y = 1 | X_0 = y)$ is equal to 1 and $\rho_{y,y} = 1$. An absorbing state is therefore recurrent!
    Let $N(y)$ be the random number of times, possibly infinite, that the Markov chain takes the value y:

$$N(y) = \sum_{n \geq 1} \mathbf{1}_{X_n = y} .$$

We then have:

$$\mathbf{P}(N(y) \geq 1 | X_0 = x) = \mathbf{P}(T_y < \infty | X_0 = x)$$
$$= \rho_{x,y} .$$

By induction:

$$\mathbf{P}(N(y) \geq m | X_0 = x) = \rho_{x,y} \rho_{y,y}^{m-1} .$$

And, since:

$$\mathbf{P}(N(y) = m | X_0 = x_0) = \mathbf{P}(N(y) \geq m | X_0 = x)$$
$$-\mathbf{P}(N(y) \geq m + 1 | X_0 = x) ,$$

we can deduce:

$$\mathbf{P}(N(y) = m | X_0 = x_0) = \rho_{x,y} \rho_{y,y}^{m-1} (1 - \rho_{y,y}) .$$

This leads to the following fundamental Theorem.

Theorem 5.2.1 *Number of visits.*

1. *If the state y is transient:*

$$\mathbf{E}(N(y)|X_0 = x) = \frac{\rho_{x,y}}{1 - \rho_{y,y}} \ .$$

   Starting from any state x, the number of visits in the state y will be (a.s.) finite.
2. *If the state y is recurrent:*
   a)

$$\mathbf{P}(N(y) = \infty|X_0 = y) = 1 \ ,$$

   and

$$\mathbf{E}(N(y)|X_0 = y) = \infty \ .$$

   Starting from the state y, the Markov chain returns an (a.s.) infinite number of times in y.
   b)

$$\mathbf{P}(N(y) = \infty|X_0 = x) = \mathbf{P}(T_y < \infty|X_0 = x)$$
$$= \rho_{x,y} \ .$$

   c) *if $\rho_{x,y} = 0$, then:*

$$\mathbf{E}(N(y)|X_0 = x) = 0 \ .$$

   If the state x does not communicate with the state y, and if the Markov chain starts from the state x, the chain cannot return in y, although this state y is recurrent.
   d) *if $\rho_{x,y} > 0$, then:*

$$\mathbf{E}(N(y)|X_0 = x) = \infty \ .$$

   If the state x communicates with the state y, and if the Markov chain starts from the state x, the chain returns an infinite number of time in y.

5.3 Subset classification

The recurrence of a state passes on to other states. In other words, recurrence classes are equivalence classes.

Lemma 5.3.1 *Recurrence.*
   Let x be a recurrence state. Assume that $\rho_{x,y} > 0$. Then the state y itself is recurrent and $\rho_{x,y} = \rho_{y,x} = 1$.

The irreducibility of a Markov chain is an essential concept.

Definition 5.3.1 *Irreducibility.*

1. *A subset C of the state space S is irreducible if $\rho_{x,y} > 0$ for all $(x,y) \in C^2$.*
2. *A Markov chain is irreducible if S is irreducible.*

Definition 5.3.2 *Closed set.*
    A subset C of the state space S is closed if $\rho_{x,y} = 0$ for all $x \in C$, $y \notin C$.

In other words, a Markov chain cannot escape from a closed subset.

Corollary 5.3.1 *Let C be a closed irreducible subset of the state space S, which states are recurrent. Then, for all $(x,y) \in C^2$:*

$$\rho_{x,y} = 1 \, ,$$
$$\mathbf{P}(N(y) = \infty | X_0 = x) = 1 \, ,$$
$$\mathbf{E}(N(y) | X_0 = x) = \infty \, .$$

Lemma 5.3.2 *Characterization of finite closed irreducible subsets.*
    Let C be a finite closed irreducible subset of the state space S. Every state of C is recurrent.

5.4 Genetical drift

5.4.1 Modeling

We will consider here an haploid modeling of stochastic reproduction, without mutation, selection or immigration. The population is therefore isolated. The genetical size of the population is assumed to be constant and equal to $2N$ genes. There exist two types of genes: type a and type A. At generation 0, there exists a given number X_0 $(0 < X_0 < 2N)$ of a-genes. The drawing of genes from a generation to the next one follows a binomial distribution. The children immediately replace the parents. Let X_n be the number of a-genes at generation n. The number of genes A is equal to $2N - X_n$. The sequence X_n is a Markov chain with transition probability:

$$\mathbf{P}(X_{n+1} = k | X_n = j) = P(j, k)$$
$$= C_{2N}^k \left(\frac{j}{2N} \right)^k \left(\frac{2N - j}{2N} \right)^{2N-k}$$
$$j, k = 0, \ldots, 2N \, .$$

The sequence X_n is an homogeneous discrete-time Markov chain with finite state space $\{0, 1, \ldots, 2N\}$.

5.4.2 Expectation

Little algebra shows that this Markov chain has a constant expectation. Firstly compute the conditional expectation:

$$\mathbf{E}(X_{n+1}|X_n = j) = \sum_{k=0}^{2N} kP(j,k)$$

$$= \sum_{k=0}^{2N} kC_{2N}^k \left(\frac{j}{2N}\right)^k \left(\frac{2N-j}{2N}\right)^{2N-k}$$

$$= j .$$

So that:

$$\mathbf{E}X_{n+1} = \mathbf{E}(\mathbf{E}(X_{n+1}|X_n))$$

$$= \sum_{j=0}^{2N} \mathbf{P}(X_n = j)\mathbf{E}(X_{n+1}|X_n = j)$$

$$= \sum_{j=0}^{2N} \mathbf{P}(X_n = j)j$$

$$= \mathbf{E}X_n$$

$$= \mathbf{E}X_0 .$$

The expectation is therefore constant and is given by the initial condition. This agrees with the Hardy-Weinberg law: the genotype frequencies is stable. If one confuses the Markov chain and its expectation, one can deduce that an isolated population does not evolve. We will see that this conclusion is wrong.

5.4.3 Asymptotical behavior

Let us study how the communication between the states of this Markov chain works. Two types of states exist.

- The states $\{1, 2, \ldots, 2N - 1\}$.
  Genes a and A coexist. Starting from any state of $\{1, 2, \ldots, 2N - 1\}$, one can go to any state of $\{1, 2, \ldots, 2N - 1\}$ in one step since the transition matrix coefficients $P(j,k)$ are positive.
- The states 0 and $2N$.
  One of the genes has disappeared. These two states are absorbing: if $X_n = 0$ (resp. $2N$) then $X_{n+p} = 0$ (resp. $2N$) for all $p \geq 0$. Moreover, any state of $\{1, 2, \ldots, 2N - 1\}$ can lead to the states 0 or $2N$ in one step.

Starting from any state of $\{1, 2, \ldots, 2N - 1\}$, one can go to the states 0 or $2N$ in one step. The states 0 and $2N$ are absorbing: the chain X_n cannot return to its initial state. The states $\{1, 2, \ldots, 2N - 1\}$ are transient. The states 0 and $2N$ are recurrent. Using Theorem 5.2.1, we see that the chain X_n only returns a finite number of time in the subset $\{1, 2, \ldots, 2N - 1\}$. It reaches therefore either the state 0, either the state $2N$ in a finite time. One of the two genes a or A disappears. This very simple modeling[3] shows that an isolated population looses its genetical variability: the Hardy-Weinberg law fails.

5.5 Invariant measure

A crucial question arises: what is the limiting behavior of the Markov chain (X_n) as $n \to +\infty$? Is there any stabilization of the chain? Some convergence to an invariant measure?

Definition 5.5.1 *Invariant measure.*
A probability π on the state space $\mathcal{S}$ is an invariant measure if:

$$\sum_{x \in \mathcal{S}} \pi(x) P(x, y) = \pi(y) \quad \forall y \in \mathcal{S} .$$

By induction, one satisfies that, if π is an invariant measure,

$$\sum_{x \in \mathcal{S}} \pi(x) P^n(x, y) = \pi(y) \quad \forall y \in \mathcal{S} .$$

If X_0 follows an invariant distribution π, so is X_n:

$$\mathbf{P}(X_n = y) = \pi(y) \quad \forall y \in \mathcal{S} .$$

Reciprocally, let us assume that the distribution of X_n does not depend on the time n. Then the initial distribution π_0 satisfies:

$$\begin{aligned}
\pi_0(y) &= \mathbf{P}(X_0 = y) \\
&= \mathbf{P}(X_1 = y) \\
&= \sum_{x \in \mathcal{S}} \pi_0(x) P(x, y) ,
\end{aligned}$$

and π_0 is an invariant measure.

---

[3] This modeling is very simple but leads to numerous applications. Usually, a population cannot be isolated during a long period. Therefore, the extinction due to the genetical drift is unlikely when the population size is huge. It follows that the larger the population size, the greater the genetical variability. The population size is usually proportional to the area. The genetical variability is roughly speaking a power function of the area (indeed there are various estimates of this power). It follows that small isolated islands have a poor genetical variability. Similarly, the amazonian deforestation will reduce the genetical variability.

Set:

$$N_n(y) = \sum_{k=1}^{n} 1_{X_k=y} \, ,$$

$$m(y) = \mathbf{E}(T_y|X_0 = y) \, .$$

$N_n(y)$ is the number of visits at y until time n, while $m(y)$ is the expectation of the return time at time n with initial condition y.

Theorem 5.5.1 *Law of large number.*
　Let y be a recurrent state. Then:

$$\lim_{n\to+\infty} \frac{N_n(y)}{n} = \frac{1_{T_y<\infty}}{m(y)} \quad (a.s.) \, ,$$

$$\lim_{n\to+\infty} \frac{\mathbf{E}(N_n(x)|X_0 = y)}{n} = \frac{\rho_{x,y}}{m(y)} \, .$$

Corollary 5.5.1
　Let C be a closed irreducible subset of the state space S with recurrent states. Then, for all $(x, y) \in C^2$:

$$\lim_{n\to+\infty} \frac{\mathbf{E}(N_n(x)|X_0 = y)}{n} = \frac{1}{m(y)} \, ,$$

and, if $X_0 \in C$, for all $y \in C$:

$$\lim_{n\to+\infty} \frac{N_n(y)}{n} = \frac{1}{m(y)} \, .$$

Now we need to classify the recurrent states.

Definition 5.5.2 *Null recurrent, positive recurrent states.*
　A recurrent state y is null recurrent if $m(y) = \infty$.
　A recurrent state y is positive recurrent if $m(y) < \infty$.

Lemma 5.5.1 *Positive recurrent states.*

1. *If x is a positive recurrent state, and if $\rho_{x,y} > 0$, then y is a positive recurrent state.*
2. *Let C be a closed irreducible finite subset of the state space S. Then all the elements of C are positive recurrent.*

5.5.1 Existence and uniqueness of the invariant measure

Lemma 5.5.2
　Let π be an invariant measure. If the state x is transient or null recurrent, then $\pi(x) = 0$.

Theorem 5.5.2 *Existence and uniqueness of the invariant measure.*
An irreducible positive recurrent chain has a unique invariant measure given by:

$$\pi(x) = \frac{1}{m(x)} \quad x \in \mathcal{S} .$$

Corollary 5.5.2
An irreducible Markov chain with a finite state space has a unique invariant measure.

Corollary 5.5.3   *Consider an irreducible, positive recurrent chain with invariant measure π. Then, for all $x \in \mathcal{S}$:*

$$\lim_{n \to +\infty} \frac{N_n(x)}{n} = \pi(x) .$$

5.5.2 Aperiodic chain

We just have seen that, for an irreducible, positive recurrent chain with an invariant measure, and for all $(x, y) \in \mathcal{S}^2$:

$$\lim_{n \to +\infty} \frac{1}{n} \sum_{k=1}^{n} P^k(x, y) = \pi(y) .$$

We will now see more powerful results for aperiodic chains. We first need to define the periodicity of a state.

Definition 5.5.3 *Periodicity of a state.*
The period d_x of a state x is defined by:

$$d_x = g.c.d.\{n \geq 1, \ P^n(x, x) > 0\} .$$

We can deduce the following properties of the period:

- If $P(x, x)$ is positive, the period of x is 1.
- Let x and y two states such that $\rho_{x,y} > 0$, then x and y have the same period.
- The states of an irreducible chain have the same period. This period is called the period of the chain.

Definition 5.5.4
An irreducible chain is called an aperiodic chain if its period is 1.

An aperiodic, irreducible, positive recurrent chain converges to its invariant measure for every initial condition.

Theorem 5.5.3 *Convergence to the invariant measure for an aperiodic chain.*
Consider an aperiodic, irreducible, positive recurrent chain with invariant measure π. Then, for all, $(x, y) \in \mathcal{S}^2$:

$$\lim_{n \to +\infty} P^n(x, y) = \pi(y) \, .$$

This introduction to Markov chains will be illustrated in the section "the domestication of pearl millet"; this section will provide with a more substantial application of Markov chains than the basic example of the genetical drift.

5.6 Continuous-time

The continuous-time framework is much more delicate than the discrete-time one. Firstly, we will give the construction of the Brownian motion and of the diffusion processes. Secondly, we will give a sketch of the approximation of a Markov chain by a diffusion process. These notions will then be re-used in the section "the domestication of pearl millet". See [43] for an elementary presentation of the following results, to [34, 32, 68] for more general results on diffusion processes and stochastic differential equations, and to [48] for questions related to the moving from discrete-time to continuous-time.

5.6.1 A construction of Brownian motion

We start by a fundamental example that illustrates what will be done later.
Let X_n, $n \geq 0$ be a Markov chain on the integers defined by the transition probability:

$$X_0 = 0 \, ,$$
$$\mathbf{P}(X_{n+1} = i + 1 | X_n = i) = \frac{1}{2} \, ,$$
$$\mathbf{P}(X_{n+1} = i - 1 | X_n = i) = \frac{1}{2} \, .$$

This Markov chain models the simplest random walk. Let ε be the one-step increment:

$$\varepsilon_n = X_n - X_{n-1} \, .$$

The ε_n are an i.i.d. sequence of r.v. with distribution:

$$\mathbf{P}(\varepsilon_n = 1) = \frac{1}{2} \, ,$$
$$\mathbf{P}(\varepsilon_n = -1) = \frac{1}{2} \, .$$

The idea is to perform a double *renormalization*, one in time, the other in space. Let us therefore define the sequence W_t^n, $n \geq 0$, $t \in [0, 1]$:

$$W_t^n = \frac{1}{\sqrt{n}} X_{[nt]}$$

$$= \frac{1}{\sqrt{n}} \sum_{k=1}^{[nt]} \varepsilon_k .$$

The time renormalization factor is $1/n$, the space renormalization factor is $1/\sqrt{n}$. We check that the W_t^n are centered. W_t^n can be written:

$$W_t^n = \sqrt{t} \frac{1}{\sqrt{nt}} \sum_{k=1}^{[nt]} \varepsilon_k .$$

The Central Limit Theorem shows that, for any given t, as $n \to +\infty$, W_t^n converges to a centered Gaussian variable with variance t. The increments of process W_t^n become, for $t' > t$:

$$W_{t'}^n - W_t^n = \sqrt{t' - t} \frac{1}{\sqrt{n(t' - t)}} \sum_{k=[nt]}^{[nt']} \varepsilon_k .$$

The Central Limit Theorem shows that, for any given (t, t'), as $n \to +\infty$, the increment $W_{t'}^n - W_t^n$ converges to a centered Gaussian variable with variance $t' - t$. Similarly we check that for $t_1 < t_2 < t_3 < t_4$, the increments $W_{t_4}^n - W_{t_3}^n$ and $W_{t_2}^n - W_{t_1}^n$ are independent. The rigorous construction of the Brownian motion requires convergence results on stochastic processes that we will omit. We can nevertheless give the main idea. The Brownian motion is defined as the limit of the process W_t^n as $n \to +\infty$. The Brownian motion W_t on $[0, 1]$ satisfies the following properties:

- $W_0 = 0$.
- $\mathbf{E} W_t = 0$.
- $\mathbf{E} W_t^2 = t$.
- $\mathbf{E}(W_{t'} - W_t)^2 = |t' - t|$.

The Brownian motion has thus been introduced as the limit of a discrete-time Markov chain with a countable state space.

5.6.2 Diffusion processes

Markov processes

This construction of the Brownian motion can be generalized. Firstly, we want to know what the limiting process of a Markov chain could be. So we introduce the diffusion processes on $\mathbb{R}$. We favor a heuristic approach rather than a rigorous and formal approach.

Let $X(t), t \geq 0$ a real-valued stochastic process. We assume that for every time $0 \leq t_1 < t_2 < \ldots < t_k$, the vector $(X(t_1), X(t_2), \ldots, X(t_k))$ has a probability density $p(t_1, x_1; t_2, x_2; \ldots; t_k, x_k)$.

Definition 5.6.1 *Markov process.*
 X is a Markov process if, for every Borelian set $B \subset \mathbb{R}$ and $t_{k+1} > t_k$:

$$\mathbf{P}(X(t_{k+1}) \in B | X(t_1) = x_1, X(t_2) = x_2, \dots, X(t_k) = x_k)$$

$$= \mathbf{P}(X(t_{k+1}) \in B | X(t_k) = x_k) .$$

This definition of continuous-time Markov processes is indeed similar to the definition of discrete-time Markov processes. It is therefore natural to introduce the transition probability $P(s, x, t, B)$:

$$P(s, x, t, B) = \mathbf{P}(X(t) \in B | X(s) = x)$$

$$= \int_B p(s, x; t, y) dy ,$$

with $s < t$. The function $p(s, x; t, y)$ is the transition probability density. It describes the infinitesimal probability for the Markov process to move from the state x to the state y.

A Markov process is stationary if its transition probability density $p(s, x; t, y)$ only depends on the difference $t - s$. We have seen that the increments $W_t - W_s$ of the Brownian motion follow a Gaussian distribution with variance $|t - s|$. The transition probability density of the Brownian motion is then:

$$p(s, x; t, y) = \frac{1}{\sqrt{2\pi(t - s)}} \exp\left(-\frac{(y - x)^2}{2(t - s)}\right) .$$

Diffusions

A Markov process with transition probability density $p(s, x; t, y)$ is a diffusion if it satisfies the three following properties:

1.

$$\lim_{t \downarrow s} \int_{|y - x| > \varepsilon} p(s, x; t, y) dy = 0 ,$$

2.

$$\lim_{t \downarrow s} \frac{1}{t - s} \int_{|y - x| < \varepsilon} (y - x) p(s, x; t, y) dy = a(s, x) ,$$

3.

$$\lim_{t \downarrow s} \frac{1}{t - s} \int_{|y - x| < \varepsilon} (y - x)^2 p(s, x; t, y) dy = b(s, x) .$$

The first condition involves the continuity (in probability) of the sample paths of the diffusion. The function $a(s, x)$, defined by the second condition, is the *drift* of the diffusion. The function $b(s, x)$, defined by the third condition, is the diffusion coefficient of the diffusion. We can deduce from these three conditions that a diffusion satisfies:

$$a(s, x) = \lim_{t \downarrow s} \frac{1}{t - s} \mathbf{E}(X(t) - X(s) | X(s) = x) , \qquad (5.2)$$

$$b(s, x) = \lim_{t \downarrow s} \frac{1}{t - s} \mathbf{E}((X(t) - X(s))^2 | X(s) = x) .$$

The functions $a(s, x)$ and $b(s, x)$ are consequently called the *infinitesimal moments* of the diffusion. For practical purposes, the relationships (5.2) characterize the functions $a(s, x)$ and $b(s, x)$. Immediate computations prove that the infinitesimal moments of the Brownian motion are $a(s, x) \equiv 0$ and $b(s, x) \equiv 1$.

Example 5.6.1 *Ornstein-Uhlenbeck diffusion.*

The Ornstein-Uhlenbeck diffusion models a randomly excited oscillator. Its transition probability density $p(s, x; t, y)$ is given by:

$$p(s, x; t, y) = \frac{1}{\sqrt{2\pi(1 - \exp(-2(t - s)))}} \exp\left(-\frac{(y - x\exp(-(t - s)))^2}{2(1 - \exp(-2(t - s)))}\right) .$$

We left as an exercise to check that its infinitesimal moments are $a(s, x) = -x$ and $b(s, x) = 2$.

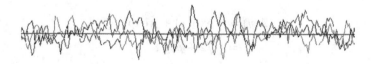

Fig. 5.1. Sample paths of Ornstein-Uhlenbeck diffusion

Itô formula

Let $X(t)$ be a diffusion with infinitesimal moments $a(x)$ and $b(x)$. Let f be a twice continuously differentiable determinist function. We want to perform the change of variables $Y(t) = f(X(t))$, and to know what the infinitesimal moments are, when they exist, of $Y(t)$. We give a heuristic computation of the change of variables formula, called Itô formula that can be rigorously done. Let us now compute the infinitesimal moments of process $Y(t)$.

$$\mathbf{E}(Y(t+h) - Y(t)|Y(t)) = \mathbf{E}(f(X(t+h)) - f(X(t))|f(X(t)))$$
$$\sim \mathbf{E}((X(t+h) - X(t))f'(X(t))$$
$$+ \frac{1}{2}(X(t+h) - X(t))^2 f''(X(t))|f(X(t)))$$
$$\sim hf'(X(t))a(X(t)) + \frac{h}{2}f''(X(t))b(X(t))$$

One mainly needs to understand the necessity of performing a second order expansion for the infinitesimal expectation of Y.

$$\mathbf{E}((Y(t+h) - Y(t))^2|Y(t)) = \mathbf{E}((f(X(t+h)) - f(X(t)))^2|f(X(t)))$$
$$\sim hf'^2(X(t))b(X(t)) .$$

Proposition 5.6.1 *Itô formula.*
The infinitesimal moments A and B of process $Y(t) = f(X(t))$ are given by:

$$A(y) = a(x)f'(x) + \frac{1}{2}b(x)f''(x) ,$$
$$B(y) = b(x)f'^2(x) ,$$

where $y = f(x)$.

Example 5.6.2
Let $X(t) = \mu + W(t)$, where $W(t)$ is a Brownian motion. Let $Y(t) = \exp(X(t))$. We obtain:

$$A(y) = \frac{1}{2}\exp(x) + \mu\exp(x)$$
$$= (1/2 + \mu)y ,$$

and

$$B(y) = \exp(2x)$$
$$= y^2 .$$

5.6.3 Diffusion approximation

It is sometimes useful to make the transition from discrete-time to continuous-time framework, and reciprocally. In the determinist framework, this corresponds for instance to the discretization of a differential equation. We then obtain a difference equation. When studying the discrete logistic model, we have seen, how difficult this transition from discrete-time to continuous-time can be. In the stochastic framework, this transition is just as difficult. One can for instance discretize a diffusion into a Markov chain, or, starting from a Markov chain, look for a diffusion that approximates this chain in a sense still to be defined. This has been outlined for the Brownian motion: a Brownian motion has been build from a random walk via a renormalization, both in time and space.

Now our aim is to generalize what has been done for the Brownian motion. Nevertheless, we will not give rigorous results[4]. We only give a heuristic approach.

Let X^N be a sequence of real-valued Markov chain. Assume that there exists a sequence of positive real number h_N converging to zero as $N \to +\infty$ and such that:

$$\mathbf{E}(X_{n+1}^N - X_n^N | X_n^N = x) \sim h_N a(x) ,$$
$$\mathbf{E}((X_{n+1}^N - X_n^N)^2 | X_n^N = x) \sim h_N b(x) .$$

A time renormalization is performed:

$$X^N(t) = X_{\left[\frac{t}{h_N}\right]}^N .$$

Under additional technical conditions, one proves the convergence in distribution of the sequence of renormalized processes $X^N(t)$ to a diffusion $X(t)$ with infinitesimal moments $a(x)$ and $b(x)$. We will now describe such an approximation of a Markov chain by a diffusion.

5.7 The domestication of pearl millet

Despite very different appearances[5], domestic plants and wild plants only differ from a few genes. They belong to the same biological species, since their hybrids are fertile. The cultivated varieties are characterized by a domestication syndrome; especially -and above all for cereals- by the sessility [6] of the corns. This sessility characteristic is spontaneously selected by the farmer. But the question of the fixation of the corns remains, since domestic and wild plants cross permanently (*e.g.* [72]).

[52] have studied the example of the African pearl millet [7] in Sahelian Africa. Until the sixties, the crop of pearl millet was characterized by:

---

[4] We will refer to rigorous results in the Bibliography given at the beginning of the chapter.

[5] For instance, teosinte (*Zea mexicana*) has a very different appearance from maize (*Zea maïs*): several stalks, ears that look like a classical graminee. Teosinte and maize were first classified in different genera. Indeed, teosinte is the wild ancestor of maize.

[6] The corns are permanently attached, and not scattered, which is essential for a plant that is to be harvested.

[7] Latin name: *Pennisetun americanum*!

1. fields of small area (several-acre-wide fields),
2. a very weak density (some plants per m^2),
3. a turn-over within several years (peanut, pearl millet, five years of fallow),
4. scattered fields,
5. omnipresence of wild pearl millet.

The plants and the ears of wild, hybrid and domestic millet can be distinguished via their morphology. There is no *in situ* conservation of hybrid or domestic plants. The genetical transfer from domestic plants to wild plants are negligible.

5.7.1 Determinist modeling

[52] consider a biallelic locus. A is the wild allele and a the domestic one. The farmer sows its fields with seeds harvested on plants of domestic genotype (aa). Thus, these seeds have a genotype aa or Aa. Let X_n and Y_n be the frequencies of these genotypes at year n. Before the blooming time, the farmer eliminates a given proportion v $(0 \leq v \leq 1)$ of hybrids. The frequencies after the flowering time become:

$$X'_n = \frac{X_n}{1 - vY_n},$$
$$Y'_n = \frac{(1-v)Y_n}{1 - vY_n}.$$

For plants (a, a), from which seeds will be harvested, the fecundation takes place:

- by autofecundation with frequency α,
- by allofecundation by the wild population with frequency m, else by the domestic plants by panmixia.

Frequencies (a, a) and (a, A) of year $n + 1$ are given by:

$$X_{n+1} = \alpha + (1 - \alpha)(1 - m)\left(X'_n + \frac{1}{2}Y'_n\right),$$
$$Y_{n+1} = (1 - \alpha)m + (1 - \alpha)(1 - m)\frac{Y'_n}{2}.$$

For Y_n:

$$Y_{n+1} = (1 - \alpha)F(m, v, Y_n),$$

with

$$F(m, v, y) = m + \frac{(1-m)(1-\alpha)y}{2(1 - vy)}.$$

An elementary study of function F shows that there exists a unique steady state Y_∞, that depends on parameters α, m and v and that is a solution to:

$$Y_\infty = (1 - \alpha)F(m, v, Y_\infty) \, ,$$

This steady state is stable. The sequence Y_n converges to Y_∞ as $n \to +\infty$. We can check that:

- Y_∞ is a decreasing function of α,
- Y_∞ is a increasing function of m,
- Y_∞ is a decreasing function of v.

We can notice that the influence of the hybrids counter-selection done by the farmer is weak. A domestic plant that mostly reproduces by autofecundation has an advantage. The means of cultivation favors the autofecundation all the more so since the contaminated flux m is strong: autofecundation is part of the domestication syndrome.

5.7.2 Stochastic modeling

Among the parameters of the determinist modeling, v has no real influence. In the sequel we consider v as constant. The parameter α is important[8], but is not intrinsically random. On the other hand, it is commonly accepted that m varies in an unpredictable way from year to year. Moreover, there exist uncertainties due to sexual reproduction. We will successively consider the effects of such uncertainties.

Fluctuation of the flux of wild pollen

Let us now consider a varying flux m_n of pollen. The m_n, $n \geq 0$, are a sequence of i.i.d. r.v. The modeling becomes:

$$Y_{n+1} = F_{m_{n+1}}(Y_n) \, ,$$

with

$$F_m(y) = (1 - \alpha) \left(m + \frac{(1 - m)(1 - v)y}{2(1 - vy)} \right) \, .$$

The sequence Y_n is therefore a Markov chain. We assume that the m_n have a common density supported by $[a, b] \subset [0, 1]$. The convergence of the determinist modeling and the increasing of the steady state (still denoted by Y_∞) in terms of m show that the chain Y_n is (a.s.) absorbed by the interval $[Y_\infty(a), Y_\infty(b)]$. This chain takes its values in a sub-interval of $\mathbb{R}$. This framework is slightly different from the previous one, but the same notions

[8] Parameter α can lead to a game theory modeling. We will not study it, though.

remain. This chain is irreducible, recurrent and aperiodic. It has a unique invariant measure. We cannot analytically give its invariant measure, which characterizes the permanent behavior of the chain. So let us note that:

$$\mathbf{E}(Y_{n+1} - Y_n | Y_n = y) = \mathbf{E}(F_{m_{n+1}}(y)) - y$$
$$= F_\mu(y) - y \; ,$$
$$\mathrm{var}(Y_{n+1} - Y_n | Y_n = y) = \mathrm{var}(F_{m_{n+1}}(y))$$
$$= (1 - \alpha)^2 (1 - \beta(y))^2 \sigma^2 \; ,$$

where μ is the expectation of the m_n, σ^2 their variance and:

$$\beta(y) = \frac{(1 - v)y}{2(1 - vy)} \; .$$

[52] introduce a new process depending on a parameter ε. This parameter is intended to converge to 0:

$$Y_{n+1}^\varepsilon = Y_n^\varepsilon + \varepsilon \left((1 - \alpha) F_{m_{n+1}}(Y_n^\varepsilon) - Y_n^\varepsilon \right) \; .$$

For $\varepsilon = 1$, this process Y_n^ε is equal to the process Y_n. The conditional moments of Y_n^ε satisfy:

$$\mathbf{E}(Y_{n+1}^\varepsilon - Y_n^\varepsilon | Y_n^\varepsilon = y) = \varepsilon(F_\mu(y) - y) \; , \tag{5.3}$$
$$\mathrm{var}(Y_{n+1}^\varepsilon - Y_n^\varepsilon | Y_n^\varepsilon = y) = \varepsilon^2 (1 - \alpha)^2 (1 - \beta(y))^2 \sigma^2 \; . \tag{5.4}$$

We can then check that, as $\varepsilon \to 0$, Y_n^ε converges to $\widehat{Y_\infty} = Y_\infty(\mu)$. Let:

$$V_n^\varepsilon = \frac{1}{\varepsilon}(Y_n^\varepsilon - \widehat{Y_\infty}) \; .$$

The conditional moments of V_n^ε are deduced from (5.3). This proves the convergence of V_n^ε to an Ornstein-Uhlenbeck diffusion, which infinitesimal moments are $-F_\mu'(\widehat{Y_\infty})y$ and $(1 - \alpha)(1 - \beta(\widehat{Y_\infty}))\sigma$. The invariant measure of the Ornstein-Uhlenbeck diffusion is a Gaussian centered distribution with variance:

$$\Sigma^2 = \sigma^2 \frac{(1 - v\widehat{Y_\infty})^2}{4(1 - v)(1 - \mu)} \; .$$

We then deduce a rough approximation[9] to the invariant measure of Y_n. It is roughly a Gaussian distribution with expectation $\widehat{Y_\infty}$ and variance Σ^2.

---

[9] This approximation is rough since it has been obtained with $\varepsilon \to 0$ though the process Y_n corresponds to $\varepsilon = 1$.

Fluctuations due to sexual reproduction

The population size is now constant and equal to N. Each individual produces a huge number of gametes and seeds. The genotypic frequencies in the seed stock are therefore equal to those of the determinist modeling. We need to take into account the randomness of the stock sampling. This modeling is similar to the one studied in the section "the genetical drift". The sequence Y_n^N is a Markov chain defined by its transition probabilities:

$$\mathbf{P}(Y_{n+1}^N = \frac{k}{N}|Y_n^N = y) = C_N^k F^k(y)(1 - F(y))^{N-k} .$$

The state space of this chain Y_n^N is finite. This is an irreducible, recurrent and aperiodic chain, except for the pathological values of the parameters. The conditional expectation is the solution of the determinist modeling. There exists an invariant measure. The distribution of the chain converges to this invariant measure. This is the equivalent of the steady state Y_∞ of the determinist modeling. Once again, this invariant measure cannot be computed. The conditional moments Y_n^N are given by:

$$\mathbf{E}(Y_{n+1}^N - Y_n^N|Y_n^N = y) = F(y) - y ,$$

$$\text{var}(Y_{n+1}^N - Y_n^N|Y_n^N = y) = \frac{F(y)(1 - F(y))}{N} .$$

We use the fact that the process $(Y_n^N)_n$ converges to Y_∞ as the population size N goes to infinity. Consider the normalized process:

$$V_n^N = \sqrt{N}(Y_n^N - Y_\infty) .$$

Its conditional moments are given by:

$$A_N(y) = \mathbf{E}(V_{n+1}^N - V_n^N|V_n^N = y)$$

$$= \sqrt{N}\left(F(Y_\infty + \frac{v}{\sqrt{N}}) - (Y_\infty + \frac{v}{\sqrt{N}})\right) ,$$

$$B_N(y) = \text{var}(V_{n+1}^N - V_n^N|V_n^N = y)$$

$$= F(Y_\infty + \frac{v}{\sqrt{N}})\left(1 - F(Y_\infty + \frac{v}{\sqrt{N}})\right) .$$

An expansion of order 1, justified since $B_N(v)$ is bounded, leads to:

$$A_N(y) = -(1 - F'(Y_\infty))y + o\left(\frac{1}{\sqrt{N}}\right) ,$$

$$B_N(y) = Y_\infty(1 - Y_\infty) + o\left(\frac{1}{\sqrt{N}}\right) .$$

In view of these moments, a natural idea is to approximate the process V_n^N by an Ornstein-Uhlenbeck diffusion. Indeed, the Ornstein-Uhlenbeck diffusion with parameter $-(1 - F'(Y_\infty))y$ and $Y_\infty(1 - Y_\infty)$, and which invariant measure is a centered Gaussian distribution with variance $\dfrac{Y_\infty(1 - Y_\infty)}{2(1 - F'(Y_\infty))}$, approximates the process V_n^N ([52]).

5.8 Bibliography

- DACUNHA-CASTELLE, D. and DUFLO, M. (1985). *Probability and Statistics.* volume 2. Springer-Verlag.
- GOUYON, P.-H., HENRY, J.-P. and ARNOULD, J. (2002). *Gene Avatars. The Neo-Darwinian Theory of Evolution.* Kluwer.
- HOEL, P., PORT, S. and STONE, J. (1972). *Introduction to Stochastic Processes.* Houghton Mifflin.
- IKEDA, N. and WATANABE, S. (1989). *Stochastic Differential Equations and Diffusion Processes.* North-Holland.
- ITÔ,K. and MCKEAN,H. (1996). *Diffusion Processes and Their Sample Paths.* Springer-Verlag.
- KARLIN, S. and TAYLOR, H. (1975). *A First Course in Stochastic Processes.* Academic Press.
- KARLIN, S. and TAYLOR, H. (1981). *A Second Course in Stochastic Processes.* Academic Press.
- OKSENDAL, B. (1995). *Stochastic Differential Equations.* Springer-Verlag.
- MEYN, S. and TWEEDIE, R. (1993). *Markov Chains and Stochastic Stability.* Springer-Verlag.
- LAREDO, C. and PERNES, J. (1988). Models for pearl millet domestication as an example of cereal domestication. *J. Theor. Biology.* **131** 289-305.
- PERNES, J. (1983). La génétique de la domestication des céréales. *La Recherche.* **146** 910-919.

5.9 Exercises

Exercise 5.9.1 *Binary Markov chain.*

Consider the Markov chain X_n defined on $\{0, 1\}$ by the transition probability:

$$\mathbf{P}(X_{n+1} = 0 | X_n = 0) = p \,,$$
$$\mathbf{P}(X_{n+1} = 1 | X_n = 0) = 1 - p \,,$$
$$\mathbf{P}(X_{n+1} = 0 | X_n = 1) = q \,,$$
$$\mathbf{P}(X_{n+1} = 1 | X_n = 1) = 1 - q \,.$$

1. Characterize this chain (recurrence, irreducibility,...).
2. Compute $\mathbf{P}(T_i = n | X_0 = 0)$, $i = 0, 1$.
3. Compute the invariant measure.

Exercise 5.9.2 *Invariant measure of the Ornstein-Uhlenbeck diffusion.*

The transition probability density of the Ornstein-Uhlenbeck diffusion $X(t)$ is given by:

$$p(s, x; t, y) = \frac{1}{\sqrt{2\pi(1 - \exp(-2(t-s)))}} \exp\left(-\frac{(y - x\exp(-(t-s)))^2}{2(1 - \exp(-2(t-s)))}\right).$$

Show that if $X(0)$ is a centered Gaussian variable with variance 1, then $X(t)$ is a centered Gaussian process with covariance $\mathbf{E}X(t)X(s) = \exp(-|t-s|)$.

Exercise 5.9.3 *Genetical drift (Section 5.4 ctd.).*

For all integer n, compute the variance of X_n. Interpret the results.

Exercise 5.9.4 *Birth and Death Markov chain.*

A random time is said to be an exponential time of parameter $\alpha > 0$ if it has a density $p(t) = \alpha e^{-\alpha t}$, $t \geq 0$.

Consider a random sequence $(X_t)_{t \geq 0}$ defined on the integers by using the following rules. Assume the sequence X_t to be in the state i at time t. Two independant exponential times D_i and B_i with parameter α_i and μ_i are run. If $D_i > B_i$, X_t jumps to $i - 1$ at time $t + D_i$. If $D_i < B_i$, X_t jumps to $i + 1$ at time $t + B_i$.

1. Let T be an exponential time. Show that, for $t, s > 0$:

$$\mathbf{P}(T > t + s | T > s) = \mathbf{P}(T > t).$$

2. Show that the sequence $(X_t)_{t \geq 0}$ is a Markov chain.
3. Show that the transition probabilities of $(X_t)_{t \geq 0}$ satisfy:

$$\mathbf{P}_h(i, i+1) = \alpha_i h + o(h),$$
$$\mathbf{P}_h(i, i-1) = \mu_i h + o(h),$$
$$\mathbf{P}_h(i, i) = 1 - (\alpha_i + \mu_i)h + o(h).$$

4. Show the backward and forward Kolmogorov equations:

$$\frac{\partial}{\partial t}\mathbf{P}_t(i, j) = \alpha_i \mathbf{P}_t(i+1, j) + \mu_i \mathbf{P}_t(i-1, j) - (\alpha_i - \mu_i)\mathbf{P}_t(i, j),$$

$$\frac{\partial}{\partial t}\mathbf{P}_t(i, j) = \alpha_{j-1}\mathbf{P}_t(i, j-1) + \mu_{j+1}\mathbf{P}_t(i, j+1) - (\alpha_j - \mu_j)\mathbf{P}_t(i, j).$$

5. Solve the forward Kolmogorov equation when $(X_t)_{t \geq 0}$ is a pure birth chain (*i.e.* $\mu_i = 0$ for all i).

Exercise 5.9.5 *Jukes-Cantor model (from [39]).*

We model the DNA sequence by a Markov chain. The state space S is a four-element set, built up with the four nucleotides A, C, G, T that form DNA. Let $p_{i,j}$ the probability that the base i mutates to become the base j. We assume that all possible nucleotides substitutions are equally likely. Let α be the rate of substitution. The transition matrix P becomes:

$$\begin{pmatrix} 1-\alpha & \alpha/3 & \alpha/3 & \alpha/3 \\ \alpha/3 & 1-\alpha & \alpha/3 & \alpha/3 \\ \alpha/3 & \alpha/3 & 1-\alpha & \alpha/3 \\ \alpha/3 & \alpha/3 & \alpha/3 & 1-\alpha \end{pmatrix}$$

1. Show that P^n is equal to:

$$\begin{pmatrix} a_n & b_n & b_n & b_n \\ b_n & a_n & b_n & b_n \\ b_n & b_n & a_n & b_n \\ b_n & b_n & b_n & a_n \end{pmatrix}$$

with

$$a_n = 1/4 + 3/4(1-4\alpha/3)^n ,$$
$$b_n = 1/4 - 1/4(1-4\alpha/3)^n .$$

2. What is the invariant measure of this chain?
3. Consider two DNA sequences having the same ancestor. The time n since this ancestor exists is unknown. Let p be the fraction of sites that differ between the two sequences. Justify the following estimate of n:

$$\widehat{n} = \frac{\log(1-4p/3)}{\log(1-4\alpha/3)} .$$

Exercise 5.9.6 *Word counts in a DNA sequence (from [73]).*

Consider a stationary Markov chain on a finite state space S, with transition matrix $P = (P(x,y))$, $(x,y) \in S^2$. Assume $P(x,y) > 0$, $\forall(x,y) \in S^2$.

1. Show that there exists a unique invariant measure μ. We can now assume that the initial distribution of the chain is μ.
2. We observe a sample path $X_1, \ldots, X_n$ of the chain. The number of occurrences of the word xy is defined by:

$$N(xy) = \sum_{i=2}^{n} 1_{\{X_{i-1}=x; X_i=y\}} .$$

Show that the likelihood $L(P, X_1, \ldots, X_n)$ of the sequence $X_1, \ldots, X_n$ is:

$$L(P, X_1, \ldots, X_n) = \mu(X_1) \prod_{(x,y)\in S^2} P(y,x)^{N(xy)} .$$

3. Deduce that the likelihood estimate of the $P(x, y)$ is:

$$\widehat{P}(x, y) = \frac{N(yx)}{\sum_{y \in S} N(yx)} .$$

(Hint: if $x/y = x'/y'$, then $x/y = x'/y' = (x + x')/(y + y')!$).

4. Let $W = w_1, \ldots, w_h$ be a word of length h. Let

$$N(W) = \sum_{i=h}^{n} 1_{\{X_{i-h+1}=w_1, \ldots, X_i=w_h\}}$$

be the number of occurrences of the word W in the sequence $X_1, \ldots, X_n$. Compute $\mathbf{E}N(W)$.

5. Propose an estimate of the invariant measure μ based on the law of large number.

6. Propose an estimate of $\mathbf{E}N(W)$.

7. Application. We model the DNA sequence by a Markov chain. The state space S is a four-element set, built up with the four nucleotides A, C, G, T that form DNA. We consider the DNA of $E.coli$, constituted of $n = 4638858$ nucleotides. We are looking for the word $Chi = GCTGGTGG$[10]. This word has been observed 499 times, though the estimate of $\mathbf{E}N(Chi)$ is 70. What should be established in order to claim that the word Chi of $E.coli$ has a significantly high frequency?

---

[10] Chi=Cross-over Hotspot Instigator.

6

Random arborescent models

6.1 Introduction

The study of family trees and especially the passing on of family names were at the origin of branching processes. In their temporal form, such processes can model a random family tree. What is more interesting is usually the size of the various generations and frequently asked questions concern the probability for the tree to come to an end and, if not, its evolution. A branching tree is a Markov chain. Nevertheless, it seems more judicious to have an independent approach to branching processes, which we will have here. Since the results are easily obtained, we included the demonstration or part of it in most of the cases. Apart from their applications to genealogy and demography, the branching processes can model numerous phenomena, for instance nuclear fission in physics. We will propose a recent application to a modern technique of duplication of DNA: the Polymerase Chain Reaction. We will then extend the previous modeling from trees to lattices: a short section is devoted to percolation.

Time branching processes have a space extension of the utmost interest if we want to model the diffusion of a population. Indeed, as we will see, the classical model of a diffusion (as studied in section 2.5) does not show satisfactory enough the phenomena of a quick diffusion such as the spreading of an epidemic or the colonization of a wasteland by an expanding population. The reaction-diffusion equations model the transfer through a fixed borderline but are unsuited for a moving borderline. Space branching processes allow a more specific model. A space branching process is composed of two ingredients: a time branching process and a law on the spreading of individuals. We will speak for the case of a super-critical time branching process without extinction, meaning that the size of population will exponentially grow. The law of dispersion allows more subtle models. Indeed, it will produce rare events called "large deviations" in the theory of probability. Such rare events cause high values of dispersion. They can be rare, but the exponential growth of the population will make them more and more numerous, though. Consequently, the space branching process will colonize the land very quickly. We will then give a recent application of such results concerning the oak tree in Europe after the last glaciation. We will explicitly propose a comparison between models via reaction-diffusion equations and space branching processes, based on numerous simulations.

6.2 Temporal branching processes

Branching processes model stochastic genealogical trees, starting, by convention, from a unique ancestor. In the simplest form, this genealogical tree is defined iteratively.

Definition 6.2.1 *Galton-Watson process*[1].

- *In the first generation $n = 0$, the unique ancestor died after a given life time and simultaneously has k children with probability p_k.*
- *At the n-th generation, all the individuals died after the same life time as the common ancestor, and each of them has simultaneously and independently from the others k children with probability p_k.*

A Galton-Watson process is then the stochastic sequence of the population size, generation after generation. The size is denoted by $Z_n, n \geq 0$. Clearly, the distribution of the Galton-Watson process is entirely defined by the discrete probability $p_k, k \geq 0$.

From now on, we will only consider non-trivial processes: there will always exist a couple of integers k, k', such that p_k and $p_{k'}$ are non-null. We left as an exercise the case $p_0 + p_1 = 1$. Indeed, this case requires a particular treatment and has not any real biological application. Moreover, we can assume from now on that:

$$\sum_{k \geq 0} k^2 \, p_k < \infty \, .$$

This condition is always satisfied in biological applications.

A Galton-Watson process can be defined as a Markov chain. Let $p_j^{*\,i}$ be the distribution obtained by the convolution of the offspring distribution:

$$p_j^{*\,i} = \sum_{k_1 + \ldots + k_i = j} p_{k_1} \cdots p_{k_i} \, .$$

---

[1] For equity, these processes should be called Bienaymé-Galton-Watson processes, since Bienaymé (1796-1878), in 1845, was the first to give a correct proof on the extinction probability; the work of Galton (1822-1911) and Watson (1827-1903) are dated back to 1874.

Definition 6.2.2 *Another (equivalent) definition of the Galton-Watson process.*

A Galton-Watson process is a Markov chain on positive integers which transition probability is given by:

$$Q(i,j) = \mathbf{P}\left\{Z_{n+1} = j \,|\, Z_n = i\right\} = \begin{cases} p_j^{*\,i} & \text{if } i \neq 0, \\ \delta_{0j} & \text{else.} \end{cases}$$

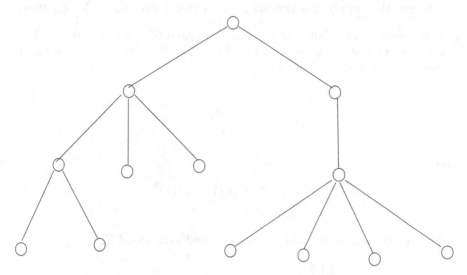

Fig. 6.1. Galton-Watson process with four generations

6.2.1 Probability generating function

The probability generating function of the process is given by:

$$\phi(s) = \sum_{k=0}^{+\infty} p_k s^k, \quad \forall s \in [0,1].$$

The probability generating function of the population size Z_n at generation n is obtained via the following Lemma.

Lemma 6.2.1

Let $\xi_i, i \geq 1$ be a sequence of i.i.d. positive r.v., with common generating function ϕ. Let Y be an r.v. on positive integers, independent from the sequence $\xi_i, i \geq 1$, with probability generating function ψ. The probability generating function of the sum $Z = \sum_{i=1}^{Y} \xi_i$ is given by $\mathbf{E}\left(s^Z\right) = \psi \circ \phi(s)$, for $s \in [0, 1]$.

Proof

$$\begin{aligned}
\mathbf{E}(s^Z) &= \mathbf{E}(\mathbf{E}(s^Z|Y)) \\
&= \mathbf{E}((\mathbf{E}(s^{\xi_1}))^Y) \\
&= \mathbf{E}(\phi(s)^Y) \\
&= \psi \circ \phi(s) .
\end{aligned}$$

We can then apply this Lemma to the random sum $Z_n = \sum_{i=1}^{Z_{n-1}} \xi_i$, where ξ_i is the number of children of the i-th individual of generation $n-1$. This leads to the probability generating function of Z_n: $\mathbf{E}\left(s^{Z_n}\right) = \phi_{Z_n}(s) = \phi_n(s)$, where $\phi_n = \phi \circ \phi \circ \ldots \circ \phi$, n times.

Let:

$$\begin{aligned}
m &= \phi'(1) \\
&= \mathbf{E}Z_1 ,
\end{aligned}$$

and

$$\begin{aligned}
\sigma^2 &= \phi''(1) + \phi'(1) - \phi'(1)^2 \\
&= \mathrm{var}(Z_1) .
\end{aligned}$$

We can recursively deduce the expectation and variance of Z_n:

$$\mathbf{E}(Z_n) = m^n ,$$

$$\mathrm{var}(Z_n) = \begin{cases} \sigma^2 m^{n-1} \frac{m^n - 1}{m-1} & \text{if } m \neq 1 , \\ n\sigma^2 & \text{else.} \end{cases}$$

6.2.2 Extinction probability

When the population size vanishes, then the future population sizes are always zero. In other words, the event $\{Z_n = 0\}$ implies $\{Z_{n+1} = 0\}$ and the sequence of events $\{Z_{n+1} = 0\}$ is increasing. The event *extinction of the population* is equal to $\{\exists n , \ Z_n = 0\}$ or $\limsup_n \{Z_n = 0\}$.

We know that:

$$\mathbf{P}\left\{\limsup_n \{Z_n = 0\}\right\} = \lim_{n \to \infty} \mathbf{P}\{Z_n = 0\} .$$

Moreover, since Z_n is a r.v. on positive integer:

$$\mathbf{P}\{Z_n = 0\} = \phi_n(0) .$$

This leads to the following Proposition.

Proposition 6.2.1 *Extinction probability.*

If the expectation m of the process is less or equal than 1, the population vanishes with probability 1.

If the expectation m of the process is greater than 1, the extinction probability q of the process is the unique solution of the equation $\phi(s) = s$, $s \in [0,1)$.

Proof

Firstly note that:

$$\phi'(s) = \sum_{k \geq 1} kp_k s^{k-1} ,$$

$$\phi''(s) = \sum_{k \geq 2} k(k-1)p_k s^{k-2} .$$

The function ϕ is increasing, and strictly convex. Note that $\phi(0) = p_0 > 0$. By induction, we check that the sequence $\phi_n(0)$ is increasing, bounded by 1. It converges to a limit q satisfying $q = \phi(q)$. Let $u > 0$ such that $u = \phi(u)$. Then $\phi(0) = p_0 < \phi(u) = u$, and, by induction, $\phi_{n+1}(0) = \phi \circ \phi_n(0) < \phi(u) = u$. So $q \leq u$ and q is the smallest positive root of the equation $s = \phi(s)$.

Since ϕ is convex, continuous with $\phi(0) > 0$ and $\phi(1) = 1$, the curve $s \to \phi(s)$ cuts the line $y = x$ in at most two points, one of them being $(1,1)$.

Assume that there exists $q \in (0,1)$ such that $q = \phi(q)$. Then $q = \lim\limits_{n \to +\infty} \phi_n(0)$ and $\phi(q) - q = 0$, $\phi(1) - 1 = 0$. Rolle's Theorem applied to $\phi(x) - x$ shows the existence of $y \in]q, 1[$ such that $\phi'(y) = 1$. Since ϕ is strictly convex, $\phi'(1) = m > 1$.

If $\phi'(1) = m \leq 1$, then $\phi'(s) < 1$ for $s < 1$. So $\int_s^1 \phi'(s)ds = 1 - \phi(s) < 1 - s$, and $\phi(s) > s$ for $s < 1$: the equation $\phi(s) = s$ has not any root on $(0,1)$.

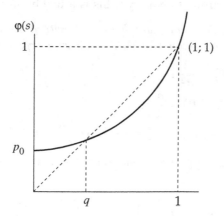

Fig. 6.2. Probability generating function: super-critical case

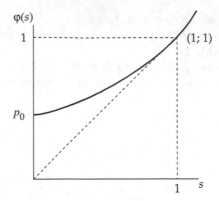

Fig. 6.3. Probability generating function: sub-critical case

6.2.3 Different types of processes

The results on expectation, variance and extinction probability lead to share the Galton-Watson processes into three groups.

- $m > 1$: super-critical processes.
- $m = 1$: critical processes.
- $m < 1$: sub-critical processes.

We will now investigate the asymptotical behavior of such processes.

Critical processes

This process vanishes with probability 1. Its expectation is equal to 1, and its variance linearly grows to infinity. This is a highly unstable process.

Proposition 6.2.2 *Asymptotical behavior of critical processes.*
Let $Z_n, n \geq 0$ be a Galton-Watson process with expectation $m = 1$. Then:

(i) $\displaystyle \lim_{n \to \infty} n\mathbf{P}\{Z_n > 0\} = \frac{2}{\sigma^2}$,

(ii) $\displaystyle \lim_{n \to \infty} \mathbf{E}\left(Z_n/n \,|\, Z_n > 0\right) = \frac{\sigma^2}{2}$,

(iii) $\displaystyle \lim_{n \to \infty} \mathbf{P}\{Z_n/n \leq u \,|\, Z_n > 0\} = 1 - \exp -\frac{2u}{\sigma^2}$, $u \geq 0$.

The proof of this Proposition relies on the following technical Lemma.

Lemma 6.2.2

$$\lim_{n \to \infty} \frac{1}{n}\left(\frac{1}{1 - \phi_n(s)} - \frac{1}{1 - s}\right) = \frac{\sigma^2}{2} ,$$

uniformly on $[0, 1[$.

Proof of Lemma 6.2.2

Let $0 \leq s < 1$. A Taylor expansion around 1 gives:

$$\phi(s) = s + \frac{\sigma^2}{2}(1-s)^2 + \varepsilon(s)(1-s)^2 \, ,$$

with $\lim\limits_{s \to 1^-} \varepsilon(s) = 0$.

$$
\begin{aligned}
\frac{1}{1-\phi(s)} - \frac{1}{1-s} &= \frac{\phi(s) - s}{(1-\phi(s))(1-s)} \\
&= \frac{\frac{\sigma^2}{2}(1-s)^2 + \varepsilon(s)(1-s)^2}{(1-\phi(s))(1-s)} \\
&= \frac{1-s}{1-\phi(s)}(\sigma^2/2 + \varepsilon(s)) \\
&= \sigma^2/2 + \delta(s) \, ,
\end{aligned}
$$

with $\lim\limits_{s \to 1^-} \delta(s) = 0$. Successive iterations lead to:

$$
\frac{1}{n}\left\{\frac{1}{1-\phi_n(s)} - \frac{1}{1-s}\right\} = \frac{1}{n}\sum_{j=0}^{n-1}\left\{\frac{1}{1-\phi \circ \phi_j(s)} - \frac{1}{1-\phi_j(s)}\right\}
$$

$$
= \sigma^2/2 + \frac{1}{n}\sum_{j=0}^{n-1} \delta \circ \phi_j(s) \, .
$$

Since $\phi_n(0) \leq \phi_n(s) \leq 1$ and since $\phi_n(0)$ converges to 1 as $n \to +\infty$, the convergence of $\phi_n(s)$ to 1 is uniform and Lemma 6.2.2 is proved.
◇

We will now use this Lemma to prove the Proposition 6.2.2.

$\mathbf{P}\{Z_n > 0\} = 1 - \phi_n(0)$. We know that $1 - \phi_n(0)$ behaves like $\dfrac{2}{n\sigma^2}$, this leads to *(i)*.

Using Bayes formula we express $\mathbf{E}(Z_n \,|\, Z_n > 0)$ in terms of $\mathbf{E}(Z_n)$:

$$\mathbf{E}(Z_n \,|\, Z_n > 0) = \frac{\mathbf{E}(Z_n)}{\mathbf{P}\{Z_n > 0\}} \, ,$$

this leads to *(ii)*.

Finally, the proof of the convergence in distribution of *(iii)* comes from the convergence of the Laplace transform of Z_n/n, conditionally to non-extinction. Indeed, using Bayes formula, we have:

$$\mathbf{E}(\exp(-tZ_n/n \,|\, Z_n > 0)) = \frac{\phi_n(\exp -t/n) - \phi_n(0)}{1 - \phi_n(0)} \, ,$$

for $t \geq 0$. Its limit is $\dfrac{1}{1 + t\sigma^2/2}$, which is the Laplace transform of $\left(1 - \exp -\dfrac{2u}{\sigma^2}\right)$.

Sub-critical processes

This process vanishes with probability 1. Its expectation and variance decays exponentially to 0. We accept the following results; their proof is not difficult, but rather tedious and without any interest for our purpose.

Theorem 6.2.1 *Asymptotical behavior of sub-critical processes.*
There exists a constant $C > 0$ such that:

$$\lim_{n \to +\infty} m^{-n} \mathbf{P}(Z_n > 0) = C .$$

Super-critical processes

This process vanishes with probability $q < 1$. Its expectation and variance grow exponentially to infinity. We prove that this process can either vanish, or explode.

Theorem 6.2.2 *Explosion of super-critical processes.*

$$\mathbf{P}(Z_n \to +\infty) = 1 - q ,$$

where q is the extinction probability of the process Z_n.

Proof
Recall that $\phi'(q) < 1$. We obtain by induction:

$$\phi_n'(q) = (\phi'(q))^n .$$

Let k and n be two integers.

$$\mathbf{P}(1 \leq Z_n \leq k) = \sum_{j=1}^{k} \mathbf{P}(Z_n = j)$$

$$\leq \sum_{j=1}^{k} \mathbf{P}(Z_n = j) \frac{j \, q^{j-1}}{q^k}$$

$$= \frac{\phi_n'(q)}{q^k}$$

$$= \frac{(\phi'(q))^n}{q^k} .$$

It follows that:

$$\sum_{n \geq 1} \mathbf{P}(1 \leq Z_n \leq k) < \infty ,$$

and we conclude with Borel-Cantelli Lemma.

The asymptotical behavior of Z_n is given by the following result.

Theorem 6.2.3 *Limiting distribution of* $\dfrac{Z_n}{m^n}$.

There exists a r.v. W with expectation equal to 1 and variance equal to $\dfrac{\sigma^2}{m(m-1)}$ *such that* $\dfrac{Z_n}{m^n}$ *converges in L^2 and a.s.to W as $n \to +\infty$. W satisfies* $\mathbf{P}(W = 0) = q$*, where q is the extinction probability.*

Proof
Set $\mathcal{F}_n = \sigma(Z_1, \ldots, Z_n)$. Then:

$$\mathbf{E}(Z_n | \mathcal{F}_{n-1}) = mZ_{n-1} .$$

The sequence $\dfrac{Z_n}{m^n}$ is a martingale and $\mathbf{E}\dfrac{Z_n^2}{m^{2n}} \leq \dfrac{\sigma^2}{m(m-1)}$. The convergence of $\dfrac{Z_n}{m^n}$ to a r.v. W follows from Theorem A.3.4[2].

Let us now show that $\mathbf{P}(W = 0) = q$. Firstly, if $Z_n \to 0$, then $W = 0$. Now consider the opposite.

$$\mathbf{P}\left(\frac{Z_n}{m^n} \to 0 | Z_1 = k \right) = \left(\mathbf{P}\left(\frac{Z_n}{m^n} \to 0 \right) \right)^k .$$

Set $\pi = \mathbf{P}(W = 0)$. We deduce:

$$\pi = \sum_{k \geq 0} \pi^k \mathbf{P}(Z_1 = k)$$
$$= \phi(\pi) .$$

Since $m > 1$, π is different from 1 and therefore $\pi = q$.
$\diamond$

For super-critical processes, the extinction probability q is equal to 0 iff $p_0 = 0$. We will admit a Central Limit Theorem and a law of the iterated logarithm in the case $p_0 = 0$, which will be useful later.

Remark 6.2.1
The Harris transform (cf. [2, Ch.I.12]) is a way of transforming a super-critical process with non-null extinction probability into a super-critical process with vanishing extinction probability. So, we can work with super-critical processes with vanishing extinction probability without any loss of generality.

Theorem 6.2.4 *Central Limit Theorem for super-critical processes.*
Assume $p_0 = 0$. Then the variable $\dfrac{Z_n - m^n W}{\sqrt{Z_n}}$ *converges in distribution, as $n \to +\infty$ to a Gaussian centered variable with variance* $\dfrac{\sigma^2}{m^2 - m}$.

---
[2] *cf.* appendix A.3.

The proof of Theorem 6.2.4 relies on the decomposition of $Z_n - m^n W$ into a sum of Z_n i.i.d. sub-random variables. Similarly, we obtain a law of the iterated logarithm.

Theorem 6.2.5 *Law of the iterated logarithm for super-critical processes. Assume $p_0 = 0$. Then:*

$$\mathbf{P}\{\omega, \ \limsup_{n \to +\infty} \frac{Z_n - m^n W}{\sqrt{Z_n}\sigma \log\log Z_n} = 1\} = 1 \ .$$

6.2.4 Maximum likelihood estimation of the expectation

The parameter m is a very important one, because the behavior of the process strongly depends on the value of such a parameter. Using the observations of the sizes of the population until the n-th generation, we can estimate the expectation m. We will estimate the expectation m using the maximum likelihood framework.

The parameter m doesn't give an exhaustive description of the distribution of the process. We need the distribution p_k, $k \geq 0$. The observation of the size is not sufficient to estimate the distribution p_k, $k \geq 0$. We will work as follows. Firstly, we will assume that we know the number of individuals having k sons for each generation from 0 to the n-th: let Z_{jk} be the number of individuals of generation j having k sons. The observations enable us to estimate the distribution of the process. Since the expectation m is simply obtained thanks to this distribution, we can deduce an estimate of m. We will then remark that this estimate of m only involves the sizes of the population. Indeed, conditionally to Z_j, the Z_{jk} have a multinomial distribution:

$$\mathbf{P}\left\{Z_{j0} = z_{j0}, Z_{j1} = z_{j1}, \ldots, Z_{jk} = z_{jk}, \ldots | Z_j = z_j\right\} =$$

$$\frac{z_j!}{\prod_{k=0}^{\infty} z_{jk}!} \prod_{k=0}^{\infty} p_k{}^{z_{jk}} \mathbf{1}\left\{\sum_{k=0}^{\infty} z_{jk} = z_j\right\} \ .$$

Since $Z_j = \sum_{k=0}^{\infty} k Z_{(j-1)\,k}$, the likelihood becomes:

$$V(\{p_k\}) = \frac{1!}{\prod z_{0k}!} \prod p_k{}^{z_{0k}} \frac{(\sum k z_{0k})!}{\prod z_{1k}!} \prod p_k{}^{z_{1k}} \ldots \frac{(\sum k z_{(n-1)\,k})!}{\prod z_{nk}!} \prod p_k{}^{z_{nk}} \ .$$

Taking the logarithm, we obtain:

$$Log V(\{p_k\}) = f(z_{00}, z_{01}, \ldots)$$

$$+ \sum_{k=0}^{\infty} \left(\sum_{j=0}^{n} z_{jk}\right) \log p_k \ , \tag{6.1}$$

where $f(z_{00}, z_{01}, \ldots)$ doesn't involve the $\{p_k\}$. Consequently we therefore will not take it into account for maximizing the likelihood.

Let $\{\pi_k, \ k \geq 0\}$ and $\{p_k, \ k \geq 0\}$ be two probability distributions. Jensen inequality shows that:

$$\sum_{k \geq 0} \pi_k Log p_k - \sum_{k \geq 0} \pi_k Log \pi_k = \sum_{k \geq 0} \pi_k Log \frac{p_k}{\pi_k}$$

$$\leq Log \sum_{k \geq 0} p_k = 0 .$$

So:

$$\sum_{k \geq 0} \pi_k Log p_k \leq \sum_{k \geq 0} \pi_k Log \pi_k . \qquad (6.2)$$

We obtain the maximum likelihood estimate of the p_k, using both (6.1) and (6.2)

$$\widehat{p}_{k,n} = \frac{\sum_{j=0}^{n} Z_{jk}}{\sum_{k=0}^{\infty} \left(\sum_{j=0}^{n} Z_{jk} \right)} .$$

The estimate of m is then:

$$\widehat{m}_n = \sum_{k=0}^{\infty} k \widehat{p}_{kn} = \frac{Y_{(n+1)} - 1}{Y_n} . \qquad (6.3)$$

As mentioned before, we can estimate the expectation m from the sizes of the population. This estimate is consistent.

Theorem 6.2.6 *Consistency of the maximum likelihood estimate.*

Conditionally to the non-extinction of the process, the estimate $\widehat{m}_n$ converges a.s. to m as $n \to +\infty$.

Proof of Theorem 6.2.6

Lemma 6.2.3

Let $a_k, k \geq 0$ be a sequence of positive real numbers such that $\lim\limits_{n \to +\infty} \sum\limits_{k=0}^{n} a_k = +\infty$. Let x_n be a real sequence converging to x as $n \to +\infty$. Then the sequence $\left(\sum\limits_{k=0}^{n} a_k \right)^{-1} \sum\limits_{k=0}^{n} a_k x_k$ converges to x as $n \to +\infty$.

$\diamond$

Since $\dfrac{Z_n}{m^n}$ converges a.s. to W:

$$\frac{\sum_{k=0}^{n} m^k \left(\frac{Z_k}{m^k} - W \right)}{\sum_{k=0}^{n} m^k} \to 0 \text{ (a.s.) .}$$

Then:

$$\frac{Y_n}{\sum_{k=0}^{n} m^k} \to W \text{ (a.s.)} .$$

It follows that: $m^{-n}Y_n \to mW/(m-1)$ a.s. Because of Theorem 6.2.3, we work conditionally to $W > 0$: this shows the consistency of $\widehat{m_n}$.

Nota-bene : See [2, 25, 36] for general results on temporal branching processes.

6.3 Polymerase Chain Reaction

The Polymerase Chain Reaction (in short PCR) is an in vitro enzymatic amplification tool (cf. [63]). Kary B. Mullis was awarded the 1993 Nobel Prize for chemistry for PCR. Starting from a small number of identical DNA sequences (the "primer"), the PCR starts at each primer and copies the sequence of that strand. Within a short time, exact replicas of the target sequence have been produced. The PCR was given a lot of media coverage with the movie "Jurassic Park": a good piece of amber was found containing an insect full of dinosaur blood, the blood cells would have to be separated from the insect's cells. The PCR replicates the DNA enough times to recover the dinosaurs. Just do it! More seriously, part of DNA sequences from a 140 millions years old weevil have been replicated by PCR.

There are three basic steps in PCR.

1. Denaturation. Firstly, the target genetic material must be denatured. The strands of its helix must be unwound and separated by heating to 90-96°C.
2. Hybridization. The second step is hybridization, in which the primers bind to their complementary bases on the now single-stranded DNA.
3. Synthesis. The third step is DNA synthesis. The polymerase reads a template strand. Then the polymerase matches it with complementary nucleotides. Two new helixes are obtained in place of the first, each composed of one of the original strands plus its newly assembled complementary strand.

When the steps work properly, the number of DNA sequences is multiplied by 2. But sometimes one step fails. The strands are not affected. Then a natural approach ([65], [66]) consists in modeling the PCR by a Galton-Watson process. Assume that there are N_0 primers. The number of primers varies between some units to thousands of them. Each of the primers is considered as the ancestor of a Galton-Watson process with probability distribution $p_1 = 1 - p$, $p_2 = p$ and $p_k = 0$ for $k \neq 1, 2$. p is the success probability of the PCR. This probability p is positive, the Galton-Watson process is super-critical, with expectation $m = 1 + p$, variance $\sigma^2 = p(1-p) = (m-1)(2-m)$ and vanishing extinction probability. The average number Z_n of DNA sequences after

n PCR is $N_0 m^n$. For biological reasons [3], it is sometimes useful to know N_0, or at least its order of magnitude. In the previous we have seen how to estimate (*cf.* estimate $\widehat{m}_n$ defined in (6.3)) the expectation from the observations of the sizes of the populations. Unfortunately, the estimation of N_0 is an ill-posed problem. Indeed, N_0 is not identifiable. Take two processes, one starting from one ancestor ($N_0 = 1$), the second starting with two ancestors ($N_0 = 2$). With probability p, the first population size Z_1 of the first process will be 2. With probability $(1 - p)^2$, the first population size Z_1 of the second process will be 2. The two probabilities associated with these two processes are not orthogonal: N_0 is not identifiable. Nevertheless, [35] proposes the following estimate of N_0:

$$\widehat{N}_{0n} = \frac{Z_n}{\widehat{m}_n} \, .$$

As $n \to +\infty$, [35], using Theorem 6.2.3 and Theorem 6.2.5, show that $\widehat{N}_{0n}$ converges (a.s.) to a variable W_{N_0}. The N_0 Galton-Watson processes starting from the N_0 ancestors have the same distribution. The variable $\widehat{N}_{0n}$ is the sum of N_0 i.i.d. variables W_i. The relative error $\dfrac{\widehat{N}_{0n} - N_0}{N_0}$ converges (a.s.) to $\dfrac{1}{N_0} \sum_{i=1}^{N_0} (W_i - 1)$ The variable $\dfrac{1}{N_0} \sum_{i=1}^{N_0} (W_i - 1)$ is centered with variance $\dfrac{1}{N_0} \dfrac{1-p}{1+p}$. This variance can be very small. For instance, if the success probability of the PCR is 80% (*i.e.* $p = 0.8$), the variance $\dfrac{1-p}{1+p}$ is about 0.11 Consequently, we can estimate the order of magnitude of N_0, even if N_0 is not identifiable.

6.4 Percolation

As a first example, let us consider a forest, in which all the trees are arranged on the vertices of a large (ideally infinite) square grid. Suppose that a tree will become infected with a disease with probability $0 < p < 1$ if one of its neighbours is infected. Will the pest spread everywhere in the forest, or will the pest stop of its own? The same question arises with forest fire. A fire starts at one or several places in the forest. When a tree burns, its neighbours will burn with probability $0 < p < 1$. Will the fire destroy all the forest or only parts of it? Let us give another example. Consider a human population, with relationships between individuals. People having relationships will be considered as neighbours. Assume that these individuals and relationships

---

[3] For instance, if the DNA comes from a sick person, we should be interested in knowing the part of contaminated DNA.

can be reasonably modeled by $\mathbb{Z}^d$, where the dimension d indeed measures the number of neighbours. An infectious disease, like AIDS, spreads into the population. Each neighbour of an infectives can be infected with probability $0 < p < 1$. Will the pest stop of its own? In the case of AIDS, the probability p can reasonably be linked with the number of safe sexual relations. How the pest does evolve when p varies? Percolation theory was first introduced by [11] as a mathematical answer to these questions. Let $\mathbb{Z}^d$ be the d-dimensional lattice and let $0 \le p \le 1$. For each $x \in \mathbb{Z}^d$, there are $2d$ edges linking x to each of its nearest neighbours. I.i.d. Bernoulli trials are done on the edges of $\mathbb{Z}^d$. An edge is open with probability p and closed otherwise, independently of the other edges. There is an open path between x and y if there exists a sequence $x_0 = x, x_1, x_2, \ldots, x_{n-1}, x_n = y$ such that the x_{i-1} and x_i are nearest neighbours and the edges $x_{i-1} - x_i$ are open. For a given point x, let $C(x)$ be the set of y such that there exists an open path from x to y. $C(x)$ is called the open cluster of x. Since our percolation model is translation invariant, the distribution of $C(x)$ is independent of x. Let us define:

$$\theta(p) = \mathbf{P}(|C(x)| = \infty) .$$

One clearly gets $\theta(0) = 0$ and $\theta(1) = 1$. The critical value of percolation on $\mathbb{Z}^d$ is defined by:

$$p_c(d) = \sup\{0 \le p \le 1, \ \theta(p) = 0\} .$$

Our first question is to know whether $|C(x)|$ is infinite or not. A second one is to know if there exists an infinite open cluster in $\mathbb{Z}^d$ somewhere. The answers to these questions are given by the following Theorem (*cf.* [24, 79]). Their proofs are postponed for exercise 6.8.7.

Theorem 6.4.1 *Percolation.*

1. *The critical value of percolation is non trivial:* $0 < p_c(d) < 1$.
2. *If $p < p_c(d)$, then $\theta(p) = 0$. If $p > p_c(d)$, then $\theta(p) > 0$.*
3. *If $p > p_c(d)$, there exists an infinite open cluster somewhere in $\mathbb{Z}^d$ with probability one.*

Moreover, one can prove (*cf.* [24]) that $p_c(2) = 1/2$. The value of $p_c(d), d > 2$ is still unknown. Very few is known on function $\theta(p)$. Let us just mention that function $\theta(p)$ is increasing and continuous for $p_c(d) < p \le 1$.

The answer to our question (will the pest/fire spread or stop on its own) is binary. Either parameter p is less than the critical value, and the pest/fire will stop; either it is more than the critical value, and the pest/fire will spread. It has important implications in terms of management. Indeed, if one has some control on the parameter p, decreasing this parameter will not have any real influence until it stays over the critical value, and suddenly, when crossing the critical value, the spread of pest/fire is stopped.

6.5 Spatial branching processes

Spatial branching processes model the spatial repartition of a random tree; this tree being described by a temporal branching process. The random tree spreads in $\mathbb{R}^k$, $k = 1, 2, 3$. In the easiest form, with $k = 1$, a spatial Galton-Watson process is described recursively.

Definition 6.5.1 *Spatial Galton-Watson process.*

- *In the first generation $n = 0$, the unique ancestor died after a given life time and simultaneously has k children with probability p_k. This ancestor is located at the origin of $\mathbb{R}$.*
- *At the n-th generation, all the individuals died after the same life time as the common ancestor, and each of them, simultaneously and independently from the others, has k children with probability p_k. The location of the k sons is distributed independently from the other individuals of the n-th generation with a dispersion distribution μ. This distribution μ is constant through time.*

The spatial Galton-Watson process is driven by the offspring distribution (p_k), $k \in \mathbb{N}$ and by the dispersion distribution μ. From now on, we assume that the Galton-Watson process is super-critical *i.e.* $\sum_{k \geq 1} k p_k > 1$.

Let us define some parameters on this process.

- The expectation of the Galton-Watson process:

$$m = \sum_{k \geq 1} k p_k \ .$$

- The log Laplace transform of the dispersion distribution:

$$L_\mu(z) = \log \int_{\mathbb{R}} \exp(tz) \mu(dt) \ .$$

- The Cramér transform of the dispersion distribution:

$$h_\mu(x) = \sup_{z \in \mathbb{R}} \left(xz - L_\mu(z) \right) \ .$$

- The "compensated" Cramér transform of the spatial Galton-Watson process:

$$h(x) = h_\mu(x) - \log m \ .$$

Since the Galton-Watson process is super-critical ($\log m > 0$), note that the function h can reach the negative values.

6.5.1 Asymptotical behavior

Let us start with an expectation approach. We have seen that the average size of the population is m^n. The location of an individual of the n-th generation is the result of a sum of n i.i.d. r.v. of distribution μ. Let $I = [a, b]$ be an interval such that $\mathbf{E}\mu \notin I$: for instance, we can chose I such that $\mathbf{E}\mu < a < b$. The order of magnitude of the probability that an individual from the n-th generation belongs to the interval $[na, nb]$ is $\exp(-nh_\mu(a))$ (see the Chernoff Theorem [4]). The expected number of individuals of generation n that belong to the interval $[na, nb]$ is of order $m^n \exp(-nh_\mu(a)) = \exp(-nh(a))$. This heuristic computation shows that the presence or absence of individuals in a given area depends on the sign of the "compensated" Cramér transform. Indeed, when the "compensated" Cramér transform is positive, the expected number of individuals of generation n being in the interval $[na, nb]$ exponentially decays to 0. When the "compensated" Cramér transform is negative, the expected number of individuals of generation n being in the interval $[na, nb]$ exponentially grows to $+\infty$. The borderline, that is the area where the "compensated" Cramér transform is vanishing, defines the colonization border of the Galton-Watson process.

Let ξ_n be the point process generated by the spatial Galton-Watson process at generation n: for every $B \subset \mathbb{R}$, $\xi_n(B)$ is the (random) number of individuals in B. The idea is to normalize this point process ξ_n through a small parameter ε. So, let us define the ε-normalization of ξ_n

$$\xi_t^\varepsilon(B) = \xi_{\left[\frac{t}{\varepsilon}\right]}\left(\frac{B}{\varepsilon}\right) ,$$

where $[x]$ is the integer part of x and $\dfrac{B}{\varepsilon}$ is the ε-dilation of the set $\{x \in \mathbb{R}/\exists y \in B, \ \varepsilon x = y\}$.

Let I_ε be the interval $[-\varepsilon^\eta, \varepsilon^\eta]$, with $0 < \eta < 1$. As $\varepsilon \to 0$, and looking for the expectation, we have:

$$\xi_t^\varepsilon(x + I_\varepsilon) = \xi_{\left[\frac{t}{\varepsilon}\right]}\left(\frac{x + I_\varepsilon}{\varepsilon}\right)$$

$$\sim \exp\left(-\frac{t}{\varepsilon}h\left(\frac{x}{t}\right)\right) ,$$

and

$$\varepsilon \log \xi_t^\varepsilon(x + I_\varepsilon) \sim -th\left(\frac{x}{t}\right) .$$

The following Theorem, due to [5], rigorously gives the results we have only outlined in expectation.

---

[4] *cf.* Appendix A.3

Theorem 6.5.1 *Asymptotical behavior of a spatial Galton-Watson process.*
Let I_ε be a sequence of centered interval which length satisfies $|I_\varepsilon| \to 0$ and $|I_\varepsilon|/\varepsilon \to \infty$ as $\varepsilon \to 0$. Then, as $\varepsilon \to 0$ we have:

- *If $h(x/t) < 0$, and conditionally to the non-extinction of the Galton-Watson process,*

$$\varepsilon \log \xi_t^\varepsilon (x + I_\varepsilon) \to -th(x/t) \quad (a.s.) .$$

- *If $h(x/t) > 0$, there exists a.s. ε_0 such that $\varepsilon < \varepsilon_0$ implies $\xi_t^\varepsilon (x + I_\varepsilon) = 0$.*

What are the speeds of colonization of the real line by a Galton-Watson process? The Theorem (6.5.1) shows that these colonization speeds are the solutions to the equation $h(c) = 0$. These colonization speeds both depend on the expectation and the Cramér transform of the dispersion distribution. The entire distribution μ is required to study the colonization speed, but only the expectation of the offspring distribution is required. In the section "the colonization of Europe by oaks", we will see an application of this Theorem.
Set:

$$\alpha(t, x) = -th(x/t) .$$

By Theorem (6.5.1), the function $\alpha(t, x)$ represents a limiting ε-log density. We can check that:

$$\frac{\partial \alpha(t, x)}{\partial t} = -h(x/t) + (x/t)h'(x/t)$$

$$\frac{\partial \alpha(t, x)}{\partial x} = -h'(x/t) .$$

Set:

$$L(z) = L_\mu(z) - \log m .$$

Standard properties of Cramér transform (*e.g.* [18]) implies the following link between $h(x)$ and $L(x)$:

$$h(x) = xh'(x) - L(h'(x)) .$$

The function $\alpha(t, x)$ thus satisfies the partial differential equation (*cf.* [53], [76]):

$$\frac{\partial \alpha(t, x)}{\partial t} = L \left(-\frac{\partial \alpha(t, x)}{\partial x} \right) .$$

Note that this partial differential equation is not a reaction-diffusion equation, but is a hyperbolic one. We therefore have the feeling that the modeling by a Galton-Watson process is not equivalent to a modeling by a reaction-diffusion equation.

6.6 The colonization of Europe by oaks

After the last glaciation, Europe was re-colonized by oaks. These oaks have remained in refuge areas in Spain, Italy and the Balkans. Oaks then spread to Northern Europe. The post-glaciation colonization by the oaks is estimated ([54]) thanks to palynological data. Palynological analyses are done with core samples. These core samples are extracted from media which are favorable to pollen conservation: lakeside sediments, peat bogs. Such analyses include carbon-dating and estimate of pollen counts. Isopollen maps have been established. A rough history of colonization can be drawn. The oaks spread at a speed of about 50 to 500 meters a year. An oak is sexually mature after about ten years. The spread speed of the oaks is high. But, most of the acorns fall near the trunk. How to explain the high speed of the oaks in the process?

6.6.1 Back to reaction-diffusion equations

The standard determinist model for diffusion (*cf.* section (2.5)) is usually based on reaction-diffusion equation. We want to model the spread speed, not to obtain a realistic model for areas already occupied by oaks. We allow a Malthusian dynamics (no limit for the growth). The standard reaction-diffusion equation is (*e.g.* [69, 70]):

$$\frac{\partial N(t,x)}{\partial t} = D\frac{\partial^2 N(t,x)}{\partial x^2} - v\frac{\partial N(t,x)}{\partial x} + rN(t,x) , \qquad (6.4)$$

$N(t,x)$ is the local density of oaks at time t, r is the reproduction rate, v is an advection parameter (for instance the wind) and D is the diffusivity. If, at time 0, the population size is M and is concentrated at the origin, the solution of this equation is:

$$N(t,x) = \frac{M}{2\sqrt{\pi Dt}} \exp\left(rt - \frac{(x-vt)^2}{4Dt}\right) .$$

Let us study the spread speed of the level set $N(t,x(t)) = ct$ as $t \to +\infty$. The two functions $x_\pm(t)$ that define the level sets satisfy:

$$\lim_{t\to\infty} \frac{x_\pm(t)}{t} = v \pm 2\sqrt{rD} .$$

The asymptotical speeds are $c_\pm = v \pm 2\sqrt{rD}$. For oaks, as indicated by the simulations in the next section, the usual parameters D, r and v lead to colonization speeds less important than the observed speeds.

6.6.2 Comparison with the stochastic model

Let us now model the oak spread by a spatial Galton-Watson process. Firstly we have to chose an offspring distribution. Only the expectation is important. We want to compare this model to the model (6.4), so we set $\log m = r$. Secondly, we have to chose a dispersion distribution μ.

Let us start with a Gaussian distribution $\mu = \mathcal{N}(\theta, \sigma^2)$. Its Cramér transform is:

$$h_\mu(x) = \frac{(x - \theta)^2}{2\sigma^2} .$$

An application of Theorem (6.5.1) shows that the colonization speeds are:

$$c_\pm = \theta \pm \sqrt{2r\sigma^2} .$$

The same colonization speeds, at least in order of magnitude, as for the reaction-diffusion model (6.4), are obtained if we assume the following equivalence between the parameters:

$$r = \log m ,$$
$$v = \theta ,$$
$$D = \sigma^2/2 .$$

Let us come back to the biological model. If we assume the dispersion to be Gaussian, we implicitly assume that the dispersion of the acorns is due to a unique factor, for instance the wind. Acorns are heavy, and wind has no real influence. The offspring distribution is almost centered and with a very small variance. But, one can observe that a minority of acorns are carried far away from the oak by animals, and especially by jays. A single jay scatters about 4600 acorns a year and at least 5 to 6% of a oak's acorns are scattered. A jay can carry several acorns but the acorns are buried separately. It seems that jays choose transition zones of vegetation to bury acorns: this is favorable for germination. The carrying distance varies from about a hundred meters to several kilometers. The maximal observed carrying distance is eight kilometers. We will therefore model the dispersion by a mixing of two distributions. The first distribution concerns most of the acorns, and is a short range distribution. The second distribution concerns a minority of acorns, but it is a long range distribution. This second distribution dramatically modifies the Cramér transform: the probability of large deviations increases. It follows that the colonization speed (*cf.* Theorem (6.5.1)) dramatically changes too. The simulations done by [82] indicates that the order of magnitude of the simulated speeds is the same as the observed speeds.

value of σ^2	mixing Gaussian (D.A.)	mixing Gaussian (S.A.)	mixing Dirac Gaussian (D.A.)	mixing Dirac Gaussian (S.A.)
10	7.94	23.38	7.28	23.35
20	14.90	46.71	14.56	46.70
30	22.01	70.06	21.84	70.05
40	29.29	93.41	29.12	93.40
50	36.53	116.75	36.39	116.75
60	43.79	140.10	43.67	140.10
70	51.05	163.45	50.95	163.45
80	58.32	186.80	58.23	186.80
90	65.59	210.15	65.51	210.15
100	72.86	233.50	72.79	233.50

value of σ^2	Gaussian (S.A.)	Laplace (S.A.)	mixing Laplace (S.A.)
10	21.55	29.54	37.1
20	43.10	59.09	75.02
30	64.66	88.63	112.51
40	86.21	118.18	150.01
50	107.76	147.72	187.50
60	129.31	177.27	225.00
70	150.86	206.81	262.50
80	172.41	236.36	299.00
90	193.97	265.90	337.49
100	215.52	295.45	374.99

The mixing has been done with rates of 95% and 5%. D.A stands for Determinist Approach (*i.e.* reaction-diffusion model with the equivalence $r = \log m$, $v = \mathbf{E}\mu$ and $D = 1/2\text{var}(\mu)$). S.A. stands for Stochastic Approach: a spatial Galton-Watson process has been used.

6.7 Bibliography

- ATHREYA, K. and NEY, P. (1972). *Branching Processes.* Springer-Verlag.
- BIGGINS, J. (1977). Chernoff's Theorem in the branching random walk. *J. Appl. Prob.* **14** 630-636.
- GRIMMETT,G. (1989). *Percolation.* Springer-Verlag.
- GUTTORP, P. (1991). *Statistical Inference for Branching Processes.* Wiley Series in Probability and Mathematical Statistics.
- JACOB, C. and PECCOUD, J. (1996). Theoretical uncertainty of measurements using quantitative polymerase. *Biophys. J.* **71** 101-108.
- JAGERS, P. (1975). *Branching Processes with Biological Applications.* Wiley Series in Probability and Statistics.

- LAREDO, C. and ROUAULT, A. (1983). Grandes déviations, dynamique de population et phénomèmes malthusiens. *Ann. Inst. Poincaré.* **19** 323-350.
- LE CORRE, V. (1997). *Organisation de la diversité génétique et histoire post-glaciaire des chênes blancs européens: approche expérimentale et par simulation.* PhD thesis, Institut national agronomique.
- OKUBO, A. (1980). *Diffusion and Ecological Problems: Mathematical Models.* volume 10. Springer-Verlag, Biomathematics.
- SCHINAZI,R. (1999). *Classical and Spatial Stochastic Processes.* Birkhäuser.

6.8 Exercises

Exercise 6.8.1 *Cell division.*

After a given period, a cell either splits into two cells with probability $0 < p < 1$, or dies without any descendant.

1. Show that this cell division can be modeled by a Galton-Watson process.
2. For which value of p is the process sub-critical, critical or super-critical? What is the extinction probability?
3. Propose an estimate of the extinction probability based on the population sizes $Z_j, 1 \leq j \leq n$. Is this estimate consistent as $n \to +\infty$?

Exercise 6.8.2 *Percolation tree and fractal spread.*

Some objects or phenomena, including leaves of plants, spreading of bacteria, soil water dynamics, are fractals and can be modeled by random trees. First consider a determinist dyadic tree: there is a unique ancestor, and each individuals has two sons, not less not more. The individuals are coded with 0 or 1 as follows:

- The ancestor is coded with 1.
- If an individual is coded with 0, its two sons are coded with 0.
- If an individual is coded with 1, its two sons are independent and follow a Bernoulli trial of parameter p, $0 < p < 1$.

1. a) Give the condition on p for having an infinite number of 1.
   b) What is the average number of 1 at the n-th generation?
   c) Propose an estimate of p based on the observations of the number of 1 over the n-th first generation.
2. Assume now $p > 1/2$. Let $0 \leq x \leq 1$. The dyadic expansion of x is a sequence of 0 and 1: there exists a natural one-to-one correspondence ϕ between the determinist dyadic tree and $[0,1]$. Let $B(1)$ be the (random) set of the branches whose all the knots are coded by 1. The set $C(1) = \phi(B(1))$ is a (random) subset of $[0,1]$. Let N_δ be the smallest number of sets of lengths at most $\delta > 0$ which can cover $C(1)$. Define, when it exists:

$$\dim_B(C(1)) = \lim_{\delta \to 0^+} \frac{\log N_\delta}{-\log \delta} .$$

$\dim_B$ is called the box dimension (e.g.[21]). Show that either $\dim_B(C(1)) = 0$ or $\dim_B(C(1)) = \log_2(2p)$ (a.s.). (nb: the Hausdorff dimension of $C(1)$ can be computed and is equal to the box dimension, cf. [27]).

3. Generalize the model to higher dimensions.

Exercise 6.8.3 *Linear fractional process.*

Consider the Galton-Watson process defined by the offspring distribution $p_k = bp^{k-1}$, $k \geq 1$ and $p_0 = \dfrac{1-b-p}{1-p}$, with $b > 0$ and $0 < p < 1$.

1. What is the expectation of the process m? What is the extinction probability q?
2. Let ϕ be the probability generating function. Show that:

$$\frac{1-p}{1-pq} = \frac{1}{m} .$$

Deduce that:

$$\frac{\phi(s) - q}{\phi(s) - 1} = \frac{1}{m} \frac{s-q}{s-1} ,$$

$$\frac{\phi_n(s) - q}{\phi_n(s) - 1} = \frac{1}{m^n} \frac{s-q}{s-1} ,$$

and give the analytical form of $\phi_n(s)$.

Exercise 6.8.4 *Total progeny.*

Let Z_n be a sub-critical Galton-Watson with expectation $m < 1$. The progeny until generation n is defined by:

$$Y_n = \sum_{k=0}^{n} Z_k .$$

The total progeny is defined by:

$$Y_\infty = \sum_{k=0}^{+\infty} Z_k .$$

1. Show that Y_∞ is (a.s.) finite.
2. Let $\psi_n(s) = \mathbf{E}(s^{Y_n})$ be the probability generating function of Y_n. Show that $\psi_n(s)$ satisfies the recursive relation:

$$\psi_n(s) = s \, \phi \circ \psi_{n-1}(s) .$$

3. Calculate $\mathbf{E}Y_n$ and $\mathrm{var}(Y_n)$.
4. Show that the probability generating function $\psi_\infty(s) = \mathbf{E}(s^{Y_\infty})$ of Y_∞ satisfies:

$$\psi_\infty(s) = s \, \phi \circ \psi_\infty(s) \,. \tag{6.5}$$

5. Show that the equation (6.5) has an unique solution.
6. Calculate the expectation and variance of Y_∞.
7. What is the distribution of Y_∞ for the cell division and for the linear fractional process?

Exercise 6.8.5 *Hitting time.*

Let Z_n be a super-critical Galton-Watson process with expectation $m > 1$. The hitting time of the level k is defined by:

$$v_k = \inf_{n \geq 0} \{Z_n \geq k\} \,,$$

with the convention $\inf \emptyset = +\infty$.

1. Show that, for $v_k < \infty$:

$$\frac{Z_{v_k - 1}}{m^{v_k}} < \frac{k}{m^{v_k}} \leq \frac{Z_{v_k}}{m^{v_k}} \,.$$

2. Let W be the (a.s.) limit of $\dfrac{Z_n}{m^n}$ as $n \to +\infty$. Show that, conditionally to the non-extinction of the process, we have:

$$\limsup_{k \to +\infty}(v_k - \log_m(k)) \leq 1 - \log_m(W) \text{ (a.s.)} \,,$$

$$\liminf_{k \to +\infty}(v_k - \log_m(k)) \geq -\log_m(W) \text{ (a.s.)} \,.$$

Exercise 6.8.6 *A continuous-time Galton-Watson process.*

A random time is said to be an exponential time of parameter $\alpha > 0$ if it has a density $p(t) = \alpha e^{-\alpha t}$, $t \geq 0$.

After an exponential time of parameter α, the first individual dies and simultaneously has k children with probability p_k. Each individual evolves like the first one and independently of the others. Let Z_t be the number of individuals at time $t \geq 0$. We assume $0 < p_0 + p_1 < 1$ and $m = \displaystyle\sum_{k \geq 0} k p_k < \infty$.

1. Let T be an exponential time. Show that, for $t, s > 0$:

$$\mathbf{P}(T > t + s | T > s) = \mathbf{P}(T > t) \,.$$

2. Let Y_k be the number of descendants of individual k, for $1 \leq k \leq Z_{n-1}$, after one unit time. Check that:

$$Z_n = \sum_{k=0}^{Z_{n-1}} Y_k .$$

3. Show that Z_n is a discrete-time Galton-Watson process.
4. Show that:

$$\frac{\partial}{\partial t}\mathbf{E}Z_t = \alpha(m-1)\mathbf{E}Z_t .$$

(Hint: use the Chapman-Kolmogorov equation).
5. Show that $\mathbf{E}Z_1 = \mathbf{E}Y_1 = e^{\alpha(m-1)}$.
6. Show that $\mathbf{P}(Z_t > 0, \forall t \geq 0) > 0$ iff $m > 1$.

Exercise 6.8.7 *Percolation (proofs of Theorem 6.4.1).*

1. Prove that the function $p \to \theta(p)$ is increasing.
2. We first work with $d = 2$. Prove that if $p < p_c(2)$ then $\theta(p) = 0$ and if $p > p_c(2)$ then $\theta(p) > 0$.
3. Let $N(n)$ the number of open self avoiding paths of length n starting at the origin of $\mathbb{Z}^2$. Prove that $\mathbf{E}N(n) \leq 4p^n 3^{n-1}$.
4. Prove that $\theta(p) \leq \mathbf{P}(N(n) \geq 1)$.
5. Prove that $\mathbf{P}(N(n) \geq 1) \leq \mathbf{E}N(n)$.
6. Prove that $p_c(2) \geq 1/3$.
7. Let $\widetilde{Z}^2$ be the dual graph of $\mathbb{Z}^2$:

$$\widetilde{Z}^2 = \{x + (1/2, 1/2),\ x \in \mathbb{Z}^2\} .$$

An edge $\tilde{a}$ of the dual $\widetilde{Z}^2$ is open (resp. closed) if it crosses an open (resp. closed) edge of $\mathbb{Z}^2$. A circuit is a path that ends at its starting point. Let A_n be the event "there exists a closed circuit of length n surrounding the origin of the dual". Show that:

$$\{|C(0)| < \infty\} = \bigcup_{n \geq 1} A_n ,$$

and that:

$$\mathbf{P}\{|C(0)| < \infty\} \leq \sum_{n \geq 1} \mathbf{P}(A_n) .$$

8. Show that:

$$\mathbf{P}\{|C(0)| < \infty\} \leq \sum_{n \geq 1} n3^{n-1}(1-p)^n .$$

9. Show that: $p_c(2) \leq \dfrac{11 + \sqrt{13}}{18}$.

10. Now consider the same problem in $\mathbb{Z}^d, d \geq 2$.
    Show that: $p_c(d) \geq p_c(d+1) > 0$.

Exercise 6.8.8 *A cellular automaton ([80], [79, Ch. V]).*

Consider a cellular automaton on $\mathbb{Z}^2$. Let p be in $(0, 1]$. At time 0, for each site $x \in \mathbb{Z}^2$, we set a 1 with probability p and a 0 with probability $1 - p$. The system evolves according to the following determinist rules. If there is a 1 at site x at time n, then we set a 1 at site x at time $n + 1$. If there is a 0 at site x at time n and if at least one neighbor in each of the orthogonal direction is a 1, then we set a 1 at site x at time $n + 1$, else we set a 0. Consider for instance the following configuration at time n:

$$1$$
$$0\ 1 \ .$$

At time $n + 1$, this configuration becomes:

$$1$$
$$1\ 1 \ .$$

We want to know what is the probability that all the sites of $\mathbb{Z}^2$ will eventually be occupied by 1.

1. Let S_k be the square whose center is 0 and whose sides has length $2k$. Let E_k be the event "each side of S_k has at least one 1 at time 0 that is not on one of the vertices of the square". Show that:

$$\mathbf{P}(E_k) = 1 - (1 - (1 - p)^{2k-1})^4 \ .$$

2. Let $E = \bigcap_{k \geq 0} E_k$. Show that $\mathbf{P}(E) > 0$.

3. Show that the event E implies that all the sites of $\mathbb{Z}^2$ will eventually be occupied by 1.

4. Show that the probability that all the sites of $\mathbb{Z}^2$ will eventually be occupied by 1 is indeed 1 (Hint: use a zero-one law).

7

Statistics

7.1 Introduction

Are statistics a mathematical model of a biological phenomenon? Or else are
they mere tools in the approach of the modeliser? Statistics undoubtedly often
play the role of an ancillary subject. However, it seemed to us that, in some
models, statistics were at the core of the problem: they do not model the
phenomenon but nothing interesting can be said about such a phenomenon
without statistics. Thus we devote a whole chapter to the subject, and not a
mere appendix.

What are statistics? Firstly, it is an estimate of parameters[1]. The most
common method of estimation lies on the notion of likelihood associated with
a model and observations. Parameters will be estimated thanks to the most
realistic values as compared to the underlying model. We will then give a
significant example coming from quantitative genetics: how to locate a major
gene- the Quantitative Trait Loci (in short QTL)- on the DNA strand with
the help of genetical markers. To do so, we will build a plausible probabilistic
model of genetic crossing-overs. Observations will be constituted via genetical
markers. The method of maximum likelihood will then give an approximate
location of the QTL on the DNA strand. Once this estimate known, it gives
rise to a series of issues:

- Is this approximate location reliable? How precise is it? How to give a
  confidence interval about such estimate?
- Is an estimate any better than any other one?
- When the presence of QTL remains hypothetical, is it possible to test its
  existence or its non-existence?

Such questions will help us study the theory of likelihood: nature of the estimate, convergence rate, Cramér-Rao bound, hypothesis tests. We will see

---

[1] Indeed we will only deal with parametric statistics: the parameters to be estimated
are finite-dimensional. We will not speak about the functional cases, *i.e.*, the case
of infinite-dimensional parameters.

there are answers to such questions but only from an asymptotical viewpoint, *i.e.* when the number of observations grows to infinity. The location of QTL is very delicate and we suggest a look at the bibliography for writings concerning its study. However, we will give a thorough example based on the life cycle of the weevil which will enable us to apply the results seen in the theoretical study of the likelihood.

7.2 Maximum Likelihood Estimate

7.2.1 Statistical model

In this book, we do not want to give a systematic lecture on statistics. We only want to provide with a sketch on standard tools of parametric statistics. More advanced results of statistics may be found in the bibliography given at the end of the chapter.

A *statistical model* is given by a measurable space $(\mathcal{X}, \mathcal{B}(\mathcal{X}))$ and a family of probability $\mathcal{P} = (\mathbf{P}_\theta)_{\theta \in \Theta}$. $\mathcal{X}$ is the space of observations. Θ is the set of possible values of parameter θ.

A statistical model is said to be dominated if there exists a sigma-finite measure f, called the dominating measure, such that, for all $\theta \in \Theta$, $\mathbf{P}_\theta$ is dominated by f. In others words, the probability $\mathbf{P}_\theta$ admits a density with respect to f. The density f will often be the Lebesgue measure.

We will only consider the case of n-sample: a n-sample is a sequence of i.i.d. r.v. $X_1, \ldots, X_n$. This is an important restriction: especially, we do not have a look to correlated observations.

As pointed out in the introduction, we only considered parametric models.

Definition 7.2.1 *Parametric model.*
The set of possible parameters Θ is a subset of $\mathbf{R}^k$.

Example 7.2.1
Consider a n-sample of Gaussian r.v. of expectation m and variance σ^2. The parameter $\theta = (m, \sigma^2) \in \mathbb{R}^2$ is of course finite-dimensional. This is a parametric model.

7.2.2 Estimation by likelihood maximum

One of the main aims of statistics is to estimate the (unknown) *true value θ_0* of the parameter, or a function $g(\theta_0)$ of the parameter. From an abstract point of view, an estimate of θ_0 (resp. $g(\theta_0)$) is a measurable function from $\mathcal{X}$ to Θ (resp. $g(\Theta)$). Of course we expect the estimate to be close to the (unknown) true value of the parameter and that this estimate will get closer and closer as the number of observations grows.

Let Θ be the set of parameters. We consider the parametric case: $\Theta \subset \mathbb{R}^k$. We observe a n-sample $X_1, \ldots, X_n$ of probability distribution $\mathbf{P}_{\theta_0}$, $\theta_0 \in \Theta$.

The true value θ_0 is unknown. The model is dominated by a measure f. Let f_θ be the density with respect to f:

$$f_\theta = \frac{d\mathbf{P}_\theta}{df} .$$

The density of the n-sample $X_1, \ldots, X_n$ is:

$$\prod_{i=1}^{n} f_{\theta_0}(x_i) .$$

The likelihood is defined by:

$$L_n(\theta, (X_i)_{1 \leq i \leq n}) = \prod_{i=1}^{n} f_\theta(X_i) .$$

This is a function of θ and of the n-sample $X_1, \ldots, X_n$.

The idea is to choose the parameter θ that makes the observations the most likely. When it exists, a maximum likelihood estimate $\hat{\theta}_n$ is defined by:

$$\hat{\theta}_n = \text{Argmax}_{\theta \in \Theta} L_n(\theta, (X_i)_{1 \leq i \leq n}) .$$

The function $\theta \to L_n(\theta, (X_i)_{1 \leq i \leq n})$ should have more than one maximum. Nevertheless, one usually speaks about *the* maximum likelihood estimate.

Example 7.2.2

Let $X_1, \ldots, X_n$ be a n-sample of Gaussian r.v. $\mathcal{N}(m_0, \sigma_0^2)$. Set $\theta_0 = (m_0, \sigma_0^2)$. The model is dominated by the Lebesgue measure. The likelihood is given by:

$$L_n(\theta, (X_i)_{1 \leq i \leq n}) = \frac{1}{(2\pi)^{n/2} \sigma^n} exp \left\{ -\frac{1}{2\sigma^2} \sum_{i=1}^{n} (X_i - m)^2 \right\} .$$

To maximize the likelihood or its logarithm $l_n(\theta, (X_i)_{1 \leq i \leq n})$ is equivalent:

$$l_n(\theta, (X_i)_{1 \leq i \leq n}) = Log(L_n(\theta))$$

$$= -\frac{n}{2} Log(2\pi) - n Log(\sigma) - \frac{1}{2\sigma^2} \sum_{i=1}^{n} (X_i - m)^2 .$$

We can then easily compute the maximum likelihood estimate:

$$\widehat{m}_n = \overline{X}$$

$$= \frac{1}{n} \sum_{i=1}^{n} X_i ,$$

$$\widehat{\sigma^2}_n = \frac{1}{n} \sum_{i=1}^{n} (X_i - \overline{X})^2 .$$

7.3 Localization of QTL

7.3.1 Some genetics

Firstly we will give a short glossary on genetics. See [86] for more sophisticated definitions and results on genetics. An individual has chromosomes in his cells that are the medium of his genetical inheritance. A chromosome is constituted of a sequence of genes, each of them standing at a permanent location on this chromosome. This location is called the locus of the gene. A gene can exist with various expressions, called alleles. At a given locus on a chromosome, there exists a unique gene "chosen" among the possible alleles. We consider the case of an eukaryote diploid species with sexual reproduction: the chromosomes appear by pairs (of homologous chromosomes), one coming from each parent. Two homologous chromosomes have the same genes, but not necessarily the same alleles. An individual is homozygote for a given gene if it has two copies of the same allele at a locus. Else, it is heterozygote.

The gametes only contain one chromosome from each pair (the fusion of two gametes leads to a complete individual). But the chromosomes of the gametes are not the copy of one of the parental homologous chromosomes. Indeed, when created, chromosomes are subject to breaking and repairing that lead to exchanges of homologous parts. So, a chromosome of a gamete is a unique mosaic, coming from the concatenation of successive fragments copied on one of the parental homologous chromosomes. This process is the recombination, leading to the genetical variability. The change-points are called crossing-overs. The farther away two loci are, the most important is the probability that a crossing-over occurs in between. If the number of crossing-overs between two loci is odd, we can say that they have recombined.

Now we need a model for this process. We assume the crossing-overs to be distributed following a Poisson process of intensity 1 (*e.g.* [26]). This implies that the number of crossing-overs between two loci follows a Poisson distribution whose parameter is the length of the interval, that the number of crossing-overs occurring in disjoint intervals are independent random variables, and that the lengths between two successive crossing-overs are exponential random variables with parameter 1. Moreover we assume the genetical transmission to be Mendelian: a gene of a gamete is a copy of one of the parental gene with probability $1/2$.

A trait specified by a unique co-dominant gene, *i.e.* such that each combination of alleles on this gene can be identified on this gene, is called a marker. A marker gives information on the genetical inheritance of an individual at a given locus.

We are interested in studying a quantitative trait y (fruit size, dairy output ...). We want to know whether the variation of this trait is due, at least partly, to a gene, whose unknown location is called Quantitative Trait Locus (in short QTL). We assume that a sequence of markers $M_i, i = 1, \ldots, n$ is known, and that their location is known without errors. The idea ([51]) is to map the QTL by looking for markers that are linked to this QTL.

In the most favorable situation, we are faced with lines (a line is a set of identical individuals that are homozygote at every locus: lines are stable by sexual reproduction). Lines are commonly used in plant improvements, but also with some animals (all the same we manage to produce males and females, which makes sexual reproduction easier...). Crosses are used to localize QTL. The simplest cross consists in crossing two lines A and B; that leads to hybrids, all identical. We then cross the hybrids with one of the parents, say A. The individuals born from this cross have the same chromosomes of type A and another, coming from the hybrid, that consists in a mosaic of types A and B. The QTL is then localized, when it exists, with this second generation.

7.3.2 QTL Likelihood

Let us fix a point on the chromosome. We want to know whether the QTL can reasonably be located at this point. The Poissonian assumption done on the crossing-overs is essential. Indeed, the QTL distribution conditionally to the state of the markers only depends on the adjacent markers M_i and M_{i+1} and does not depend on the other markers because of the independence of the jumps of the Poisson process. Let Δ be the distance between two adjacent markers and let r be the distance between the expected location of the QTL from the left-sided marker M_i. We need to know the probability that the allele of the QTL is A or B knowing that the adjacent markers are A or B. For instance, let us compute the probability that the QTL is A knowing that the adjacent markers are A and A. The adjacent markers are A and A: the number of crossing-overs on the interval $[M_i, M_{i+1}]$ is even. The QTL is A: the number of crossing-overs on each sub-intervals $[M_i, QTL]$ and $[QTL, M_{i+1}]$ is even. The probability of having an even number of crossing-overs on the interval $[M_i, QTL]$ is

$$\sum_{n \geq 0} \exp(-r) \frac{r^{2n}}{(2n)!} = 1/2(1 + \exp(-2r)) .$$

The Bayes formula then gives:

$$\mathbf{P}(QTL = A | M_i = A \ \& \ M_{i+1} = A)$$

$$= \frac{1 + \exp(-2r) + \exp(-2(\Delta - r)) + \exp(-2\Delta)}{2(1 + \exp(-2\Delta))} .$$

We now assume that the qualitative observed trait follows a Gaussian distribution with expectation μ_A and variance σ^2 if the QTL is A and a Gaussian distribution with expectation μ_B and variance σ^2 if the QTL is B. We observe a p-sample of individuals where the qualitative trait is observed. The likelihood of this p-sample is given by:

$$L_p(y) =$$

$$\prod_{i=1}^{p} \phi(\mu_A, \sigma^2) \mathbf{P}(QTL = A | M_i, M_{i+1}) \phi(\mu_B, \sigma^2) \mathbf{P}(QTL = B | M_i, M_{i+1}),$$

where ϕ is the Gaussian density.

This likelihood can be computed at any point on the chromosome. The estimate of the QTL location is the point that maximizes the likelihood.

Several questions then arise. Can we trust this location? Is our likelihood estimate the best estimate? Can we test the existence or non-existence of this QTL[2]? We will answer these questions in the following section. Let us just indicate that the answer is usually only *asymptotical*. In the QTL framework, the answer is diasymptotical ([51]): as the number p of observed individuals increases and as the distance between two adjacent markers decreases.

7.4 Asymptotical study of the likelihood

7.4.1 Consistency of the maximum likelihood estimate

The first expected property of an estimate is that this estimate converges to the true value of the parameter as the number of observations grows.

Consider a n-sample with likelihood $L_n(\theta, (X_i)_{1 \leq i \leq n})$. Using the law of large numbers, we can prove that $-\frac{1}{n} \log L_n(\theta, (X_i)_{1 \leq i \leq n})$ converges to $\mathbf{K}(\mathbf{P}_{\theta_0}, \mathbf{P}_\theta)$, where $\mathbf{K}(\mathbf{P}_{\theta_0}, \mathbf{P}_\theta)$ is the Kullback information of the distributions $\mathbf{P}_{\theta_0}$ and $\mathbf{P}_\theta$:

$$\mathbf{K}(\mathbf{P}_{\theta_0}, \mathbf{P}_\theta) = \int \frac{d\mathbf{P}_{\theta_0}}{df} \log \frac{\frac{d\mathbf{P}_{\theta_0}}{df}}{\frac{d\mathbf{P}_\theta}{df}} df$$

$$= \int f_{\theta_0} \log \frac{f_{\theta_0}}{f_\theta} .$$

The consistency of the likelihood estimate is given by the following results (*cf.* [19]).

---

[2] This question of testing the existence can be crucial. Indeed, the likelihood is a continuous function and still has a maximum. So, there always exists a likelihood estimate for the QTL, but this does not involve the existence of a QTL. We do not want to go into details (see [38]), but we have to point out that some excesses may occur. For instance, when using the QTL method, or any similar method of inverse genetic, one should wonder whether the gene of criminality or homosexuality exists. The existence of an estimate of the location of such a gene will not necessarily imply the existence of this gene.

Theorem 7.4.1 *Consistency of the likelihood estimate .*

Consider a n-sample with distribution $\mathbf{P}_{\theta_0}$*. Assume that the parametric model* $\mathbf{P}_\theta, \theta \in \Theta$ *is dominated by a sigma-finite measure* f*. Assume that the model is identifiable:* $\mathbf{K}(\mathbf{P}_{\theta_0}, \mathbf{P}_\theta) = 0$ *iff* $\theta = \theta_0$*. Assume that the set* Θ *of parameters is compact. Let* $f_\theta(x) = \dfrac{d\mathbf{P}_\theta}{df}$ *. Assume that the family of functions* $\theta \to Log f_\theta(x)$ *is equicontinuous. Then every sequence of likelihood estimates converges in* $\mathbf{P}_{\theta_0}$*-probability to* θ_0 *as* $n \to +\infty$*.*

For practical purposes, one first needs to verify that the model is identifiable. Indeed this is a sine qua non condition. Then, one needs to check the equicontinuity of the family of the log-density $Log f_\theta$.

7.4.2 Rate of convergence of estimates

We have just seen general results (*cf.* Theorem 7.4.1) that ensure the consistency of likelihood estimates. Now the following question arises: what is the rate of convergence of the estimates to the true value of the parameter? We first need to give a precise meaning to the heuristic idea of rate of convergence. The comparison of these rates then allows to rank the estimates. Usually (but this is not the only way), the rate of convergence is measured through the mean square error of the estimate. We will see that there exists a lower bound for the mean square error of estimates, and that, for regular parametric models, this bound is asymptotically reached by the likelihood estimate.

Definition 7.4.1 *Regular model.*

Consider a parametric model $\mathbf{P}_{\theta_0}, \theta_0 \in \Theta \subset \mathbb{R}^k$ *dominated by a sigma-finite measure* f*. This model is regular at point* θ_0 *if:*

1. θ_0 *is an interior point of* Θ*.*
2. *The likelihood function* $f_\theta = \dfrac{d\mathbf{P}_\theta}{df}$ *is twice continuously differentiable in a neighborhood of* θ_0*.*
3. *gradLog f_θ is a centered squared integrable r.v. for* $\mathbf{P}_{\theta_0}$*. Let us define:*

$$I_{i,j}(\theta_0) = \mathbf{E}_{\theta_0}\left(\frac{\partial}{\partial\theta_i}Log f_{\theta_0}\frac{\partial}{\partial\theta_j}Log f_{\theta_0}\right)$$

$$= -\mathbf{E}_{\theta_0}\left(\frac{\partial^2}{\partial\theta_i\partial\theta_j}Log f_{\theta_0}\right) .$$

The matrix $I(\theta_0) = (I_{i,j}(\theta_0))_{1\leq i,j\leq k}$ *is called the Fisher information matrix at point* θ_0*.*
4. $I(\theta_0)$ *is invertible.*

The following Theorem gives a lower bound for the rate of convergence of estimates.

Theorem 7.4.2 *Cramér-Rao inequality.*
Consider a regular model at point θ_0 and let $\tilde{\theta}_n$ be an estimate of θ_0.
Assume $grad\mathbf{E}(Log f_\theta \tilde{\theta}_n) = \mathbf{E}grad(Log f_\theta \tilde{\theta}_n)$. Then:

$$\mathbf{E}_{\theta_0}(\tilde{\theta}_n - \mathbf{E}_{\theta_0}\tilde{\theta}_n)^2 \geq {}^t grad\mathbf{E}_{\theta_0}\tilde{\theta}_n \frac{I^{-1}(\theta_0)}{n} grad\mathbf{E}_{\theta_0}\tilde{\theta}_n .$$

Essentially, Cramér-Rao inequality tells us that, for regular models, we cannot estimate the parameters with a greater rate than $\sqrt{n}$.

Definition 7.4.2 *Unbiased estimate.*
An estimate $\tilde{\theta}_n$ is unbiased if its expectation is equal to the true value of the parameter: $\mathbf{E}_{\theta_0}(\tilde{\theta}_n) = \theta_0$.

Definition 7.4.3 *Efficient estimate.*
An unbiased estimate is efficient if its rate of convergence reaches the Cramér-Rao inequality.

In the one-dimensional case, the Cramér-Rao bound for an unbiased estimate is $\mathbf{E}_{\theta_0}(\tilde{\theta}_n - \theta_0)^2 \geq \dfrac{I^{-1}(\theta_0)}{n}$. As a matter of fact, this is a lower bound of the variance of estimates.

Theorem 7.4.3 *Asymptotical efficiency.*
Consider a regular model at point θ_0, with likelihood f_{θ_0} and Fisher information $I(\theta_0)$. Assume the existence of a neighborhood $\mathcal{V}_0$ of θ_0 and of a squared integrable r.v. h such that for all $\theta \in \mathcal{V}_0$:

$$\left| \frac{\partial^2}{\partial \theta_i \partial \theta_j} Log f_\theta(x) \right| \leq h(x) .$$

If $\hat{\theta}_n$ is a consistent likelihood estimate, $\sqrt{n}(\hat{\theta}_n - \theta_0)$ converges in distribution to a centered Gaussian variable $\mathcal{N}(0, I^{-1}(\theta_0))$ as $n \to +\infty$.

We say that the likelihood estimate is *asymptotically efficient* . Let us recall that Theorem 7.4.3 leads to asymptotical confidence areas. Let R be a subset of $\mathbf{R}^k$. As $n \to +\infty$, $\mathbf{P}(\sqrt{n}(\hat{\theta}_n - \theta_0) \in R)$ converges to $\int_R \Phi(I^{-1}(\theta_0))$, where $\Phi(I^{-1}(\theta_0))$ is the density probability of a centered Gaussian variable of variance $I^{-1}(\theta_0)$. This leads to confidence areas for θ_0.

The following Lemma leads to the estimate of a function $g(\theta_0)$ from the estimate of θ_0.

Lemma 7.4.1
Assume the same conditions as for Theorem 7.4.3. Let g be a twice continuous differentiable function in a neighborhood of θ_0, which second order derivatives are bounded on this neighborhood. Let $\mathbf{J}g$ be the Jacobian matrix of g. Then, as $n \to +\infty$, $\sqrt{n}(g(\hat{\theta}_n) - g(\theta_0))$ converges in distribution to a centered Gaussian variable $\mathcal{N}(0, \mathbf{J}g(\theta_0)I^{-1}(\theta_0)\mathbf{J}g(\theta_0)^t)$.

7.4.3 Likelihood ratio test

We now investigate the question of testing the value of an (unknown) parameter. The aim is to decide whether the parameter belongs to a given region or not. The true parameter θ_0 belongs to a subset $\Theta \subset \mathbb{R}^k$. A (binary) test is a partition of Θ into two subsets Θ_0 and Θ_1. The subsets Θ_0 and Θ_1 are called the null hypothesis H_0 and the alternative hypothesis H_1. In view of the observations, the question is now to decide whether the true parameter θ_0 belongs to Θ_0 or Θ_1. Let us first begin with a rather abstract definition of tests. Let $X = (X_1, \ldots, X_n)$ be the observations. A test of H_0 against H_1 is defined by an acceptance region $\mathcal{R}$ such that:

- if $X \in \mathcal{R}$, H_0 is rejected,
- if $X \notin \mathcal{R}$, H_0 is accepted.

The level of the test, or error of first kind, is defined by:

$$\alpha = \sup_{\Theta_0} \mathbf{P}_{\theta_0}(X \in \mathcal{R}) .$$

α controls the probability of rejecting H_0 incorrectly.

The power of the test is defined by:

$$\sup_{\Theta_1} \mathbf{P}_{\theta_0}(X \in \mathcal{R}) .$$

The error of second kind is 1 minus the power. It controls the probability of incorrectly accepting H_0. For a given level α, one looks for a test with a maximal power. We will not investigate this question in this book. At this stage, the two hypothesis H_0 and H_1 have not any symmetric roles, not any more.

Theorem 7.4.4 *Likelihood ratio.*
Consider a n-sample satisfying assumptions of Theorem 7.4.3. Let $l_n(\theta, (X_i)_{1 \leq i \leq n})$ be the log-likelihood. Let $\widehat{\theta}_n$ be a likelihood estimate. Then, $l_n(\widehat{\theta}_n, (X_i)_{1 \leq i \leq n}) - l_n(\theta_0, (X_i)_{1 \leq i \leq n})$ converges in distribution, as $n \to +\infty$, to $\frac{1}{2}\chi^2(k)$, where $\chi^2(k)$ is a Chi-square distribution with k degrees of freedom.

The likelihood ratio test is built from this Theorem. This test is of course asymptotical. If one wants to test $H_0 = "\theta_0 = \theta^\star"$ against $H_1 = "\theta_0 \neq \theta^\star"$, for a given known value $\theta^\star$, the acceptance region of H_0 with asymptotical level α is:

$$l_n(\widehat{\theta}_n, (X_i)_{1 \leq i \leq n}) - l_n(\theta^\star, (X_i)_{1 \leq i \leq n}) \leq \frac{1}{2}\chi_\alpha^2(k) .$$

7.5 The weevil life

Now we will present an example with "real world data". This example requires standard results on regression models, and we refer to [31] for more sophisticated results.

We will describe and model some of the experiments done in the forties by S. Utida in order to study the dynamics of a weevil population *Callosobruchus chinensis* (*cf.* [75]). These experiments emphasize the effect of the population density as an intrinsic regulation mechanism of a population growth.

7.5.1 Some elements about the weevil life cycle

A weevil mainly eats stock corns. The *sex ratio* is 1:1. The mating occurs in the day following the emergence of adulthood (transition from worm to adult stage). The laying period is about one week, the exact duration depends on humidity and air pressure. Each female lays one egg a time. The egg sticks to the corn surface thanks to a gel secreted by the female. During the hatching, the individual comes into the corn; it pierces the corn surface.

7.5.2 Experimental device and results

The experimenter keeps the enclosures with constant temperature and humidity. He puts a given quantity of corns (about 20 grams) and a given number of adults couples, being just after the emergence of adulthood. He repeats the experiments for a variable number of couples. He counts the number of adults coming from the initial couples after a complete life cycle.

The results of two of these experiments are given in figure 7.1.

7.5.3 Model

For a given number x of couples of first generation, we get a random number Y of second generation adults. Y is written as a sum of its expectation, function of x, and of a centered r.v. ε: $Y = f(x) + \varepsilon$. Firstly we want to precise the shape of f.

We assume that the female potential for reproduction is equal to r_0, and that it decreases of a factor k ($k < 1$) each time she meets a female. If she meets k females, we assume the potential for reproduction to be $r_0 k^p$. In order to evaluate the average number of females met by a given female, we use the following statistical model. The weevil females spread on the corns are modeled by a Poisson point process on $\mathbb{R}^2$.

Let us recall that, for an homogeneous Poisson point process on $\mathbb{R}^2$ with parameter θ, the number of points lying in a subset B of $\mathbb{R}^2$ is a Poisson r.v. with parameter $\theta \lambda(B)$, where $\lambda(B)$ is the Lebesgue measure of B. Let $N(B)$ be this random number of points:

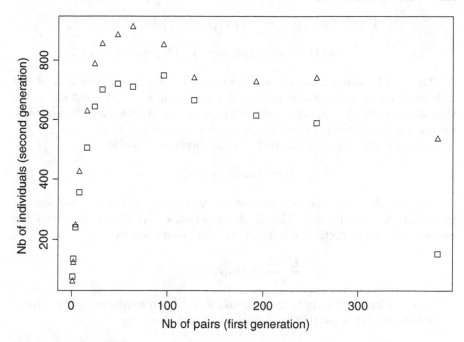

Fig. 7.1. Utida's example : graph of the observations. Δ: first experiment; $\square$: second experiment.

$$\mathbf{P}\left\{N(B) = m\right\} = \exp(-\theta\,\lambda(B))\frac{(\theta\,\lambda(B))^m}{m!}\ .$$

Moreover, if B and C are two disjoint subsets of $\mathbb{R}^2$, $N(B)$ and $N(C)$ are two independent Poisson r.v. with parameters $\theta\,\lambda(B)$ and $\theta\,\lambda(C)$.

We assume that the potential for reproduction of a female localized in u, is not weakened if she has the use on an area contained in a circle $S(u,\rho)$, centered in u, with radius ρ. The number of females in competition is given by the number of points lying in the same circle centered in u and with radius 2ρ (Two circles with same radius are disjoint iff the distance between their centers is less than 2ρ.). Therefore, we assume that the potential for reproduction r_u of a female localized in u is equal to $r_0 k^{(N[S(u,2\rho)]-1)}$. Its expectation is:

$$\mathbf{E}\left(r_u\right) = r_0 \sum_{m=0}^{\infty} k^m \mathbf{P}\left\{N\left[S(u,2\rho)\right] = m+1\,|N[\{u\}] = 1\right\}\ .$$

But,

$$\mathbf{P}\left\{N\left[S(u,2\rho)\right] = m+1\,|N[\{u\}] = 1\right\} = \mathbf{P}\left\{N\left[S(u,2\rho)\right] = m\right\}$$

since we can consider, even if it is not completely correct, that:

$$\mathbf{P}\{N[S(u,2\rho)] = m+1 \,|\, N[\{u\}] = 1\} =$$

$$\lim_{\eta \to 0} \mathbf{P}\{N[S(u,2\rho)] = m+1 \,|\, N[S(u,\eta)] = 1\} .$$

This result comes from an application of the Bayes formula and of the independence property of the number of points lying in two disjoint areas. We then obtain: $\mathbf{E}(r_u) = r_0 \exp(-\theta s(1-k))$, with $s = \lambda(S(u,2\rho))$. θs is clearly proportional to the number of females of first generation x.

A reasonable choice of function $f(x)$ is therefore given by:

$$f(x) = bx \exp(-cx) .$$

We have done the assumption that every meeting of females weakens the potential for reproduction. This is also restrictive. We decide to model the potential for reproduction of a female of first generation by

$$\frac{\mathbf{E}(Y)}{x} = b \exp(-cx^a) ,$$

where $a < 1$ is a softening parameter and $a > 1$ is an hardening parameter.

Alternative choices, that we will justify later, are:

$$\frac{\mathbf{E}(Y)}{x} = \frac{b}{(1+cx)^a} ,$$

or

$$f(x) = \frac{bx}{(1+cx)^a} .$$

7.5.4 Statistical model

Firstly we work with the following model for the regression function f:

$$Y = bx \exp(-cx^a) + \varepsilon ,$$

where

- Y is the observed size of adults of second generation;
- x is the (known) number of couples of first generation;
- $bx \exp(-cx^a)$ is the expectation of Y, it is called the regression function;
- The errors are modeled by ε. There are several causes of errors: measurement errors, variability of the parameters, ... We assume the variance of ε to be finite.
- a, b, c and σ^2 are the unknown parameters of the model.

Such a model $Y = f(x,\beta) + \varepsilon$, is called a regression model.

7.5.5 Data analysis

Now we can analyze the Utida's data. The aim is still to study the influence of the population density on its fertility. We reproduce here some of the data we will study.

	x	Y	exp.	r
1	1	77.5	I	6
2	2	136.2	I	6
3	4	240.4	I	5
4	8	356.0	I	5
5	16	505.6	I	4
6	24	643.2	I	4
7	32	700.8	I	3
8	48	720.0	I	3
9	64	710.4	I	3
10	96	748.8	I	2
11	128	666.0	I	1
12	192	614.4	I	3
13	256	588.8	I	3
14	384	153.6	I	2
15	1	65.2	IV	10
16	2	126.6	IV	10
17	4	250.0	IV	10
18	8	428.0	IV	10
19	16	630.4	IV	10
20	24	789.6	IV	10
21	32	857.6	IV	10
22	48	888.0	IV	10
23	64	915.2	IV	10
24	96	854.4	IV	10
25	128	742.4	IV	10
26	192	729.6	IV	10
27	256	742.4	IV	10
28	384	537.6	IV	3

Two experiments are summed up in this table. These experiments have been lead with different conditions of temperature and humidity (see column exp.). The number of couples of first generation is written in column x. The number of repetitions done with the same number of couples of first generation is written in column r. The expectation of the number of individuals of second generation coming from the same number of couples of first generation is written in column Y.

Our model is the following:

$$Y_i = b^I \exp\left(-c^I x_i^{a^I}\right) + \varepsilon_i/\sqrt{r_i} \quad i = 1, 2, \ldots, 14 \; ;$$
$$Y_i = b^{IV} \exp\left(-c^{IV} x_i^{a^{IV}}\right) + \varepsilon_i/\sqrt{r_i} \; i = 15, \ldots, 28 \; .$$

We assume the ε_i to be centered Gaussian r.v.'s of unknown variance σ_0^2.

Set $\beta^I = \left(a^I, b^I, c^I\right)^T$ and $\beta^{IV} = \left(a^{IV}, b^{IV}, c^{IV}\right)^T$. Denote by $f(x, \beta)$ the regression function. The log-likelihood becomes:

$$\log \ell \left(Y^{(28)} \; ; \; \beta^I, \beta^{IV}, \sigma^2\right) = -14 \log 2\pi - 14 \log \sigma^2 - \frac{1}{2} \sum_{i=1}^{28} r_i$$

$$- \frac{1}{2\sigma^2} \left(\sum_{i=1}^{14} r_i \left(Y_i - f(x_i, \beta^I)\right)^2 + \sum_{i=15}^{28} r_i \left(Y_i - f(x_i, \beta^{IV})\right)^2\right) .$$

The estimates of the parameters are the following:

	Value	Standard deviation
a_I	0,4396142	0,06661562
b_I	107,9336619	36,11258990
c_I	0,3540468	0,15559931
a_IV	0,3641156	0,03266205
b_IV	187,7486624	43,88373682
c_IV	0,5704058	0,13092695

```
sigma**2 : 16933

log likelihood :-152,306
```

We assume that a_I=a_IV and c_I=c_IV. So the estimates of the regression function are:

	Value	Standard deviation
a_I	0,3797325	0,03022036
b_I	141,2225300	28,22369967
c_I	0,5156475	0,10788360
a_IV	0,3797325	0,03022036
b_IV	172,5173688	34,66189179
c_IV	0,5156475	0,10788360

```
sigma**2 : 18041

log likelihood : -153,188
```

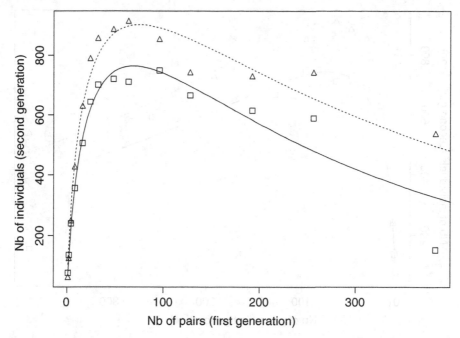

Fig. 7.2. Utida's example: data are fitted by two curves. $\triangle$: first experiment; $\square$: second experiment.

Let us choose an asymptotical level of $0,05$. The test statistics $S_{RV} = 2 \times (-152,306 + 153,188) = 1,764$ has to be compared with $v_{0,95} = 5.99$ for $\chi^2(2)$ distribution with two degrees of freedom. We accept such hypothesis.

Parameter b is indeed influenced by the experimental conditions. For this purpose we test the hypothesis $\Delta\theta_0 = (0\;0\;0)^T$, with

$$\Delta = \begin{pmatrix} 1\;0\;0\;-1\;\;\;0\;\;\;0\;0 \\ 0\;1\;0\;\;\;0\;-1\;\;\;0\;0 \\ 0\;0\;1\;\;\;0\;\;\;0\;-1\;0 \end{pmatrix}.$$

The log likelihood is $162,064$, and the test statistics $S_{RV} = 2 \times (-152,306 + 162,064) = 19,516$ has to be compared, with the same asymptotical level of 0.05, to $v_{0,95} = 7.81$ with a $\chi^2(3)$ distribution with three degrees of freedom. We reject such hypothesis (equality of the curves).

7.6 Bibliography

- DACUNHA-CASTELLE, D. and DUFLO, M. (1985). *Probability and Statistics.* Volume 1. Springer.
- DACUNHA-CASTELLE, D. and DUFLO, M. (1985). *Probability and Statistics.* Volume 2. Springer.

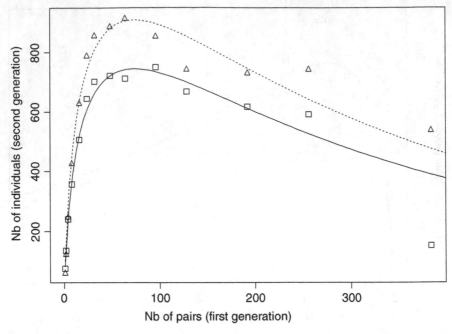

Fig. 7.3. Utida's example : the data are fitted by regression models with the same parameters a and c.

- LANDER, E. and BOTSTEIN, D. (1989). Mapping Mendelian factors underlying quantitative traits using RFLP linkage maps. *Genetics.* **121** 185-199.
- ROYAMA, T. (1992). *Analytical Population Dynamics.* Chapman & Hall.

7.7 Exercises

Exercise 7.7.1 *Mendel's peas (e.g. [17]).*

Since phenotypes are governed by dominant genes, we expect the following proportion for the four types of peas:

- Round and Yellow: 9/16,
- Round and green: 3/16,
- Angular and Yellow: 3/16,
- Angular and green: 1/16.

Now, Mendel obtained the following data by his experiment:

Phenotype	Observed frequency
Round and Yellow	315
Round and Green	108
Angular and Yellow	101
Angular and Green	32

Do you accept Mendel's law?

Exercise 7.7.2 *Student and Fisher-Snedecor tests.*

Let $X_1, X_2, \ldots, X_n$ be a sequence of i.i.d. Gaussian r.v. of unknown expectation m and unknown variance σ^2. Let $m^\star$ and $\sigma^\star$ be two given values.

1. Propose and study a test of $H_0 =$ "$m \leq m^\star$ " against $H_1 =$"$m > m^\star$" (Hint: use Theorem A.4.1).
2. Propose and study a test of $H_0 =$ "$\sigma \leq \sigma^\star$" against $H_1 =$"$\sigma > \sigma^\star$" (Hint: use Theorem A.4.1).

Exercise 7.7.3 *Comparison of the expectations in a Gaussian model.*

Let $X_1, X_2, \ldots, X_n$ be a sequence of Gaussian i.i.d. r.v. with unknown expectation m_1 and unknown variance σ^2 and let $Y_1, Y_2, \ldots, Y_m$ be a sequence of Gaussian i.i.d. r.v. with unknown expectation m_2 and unknown variance σ^2. We assume the r.v. $X_1, X_2, \ldots, X_n$ and $Y_1, Y_2, \ldots, Y_m$ to be independent. Set:

$$\overline{X} = \frac{1}{n} \sum_{i=1}^{n} X_i , \quad \overline{Y} = \frac{1}{m} \sum_{i=1}^{m} Y_i ,$$

$$S_X^2 = \frac{1}{n-1} \sum_{i=1}^{n} (X_i - \overline{X})^2 , \quad S_Y^2 = \frac{1}{m-1} \sum_{i=1}^{m} (Y_i - \overline{Y})^2 .$$

1. Let:

$$Z = \sqrt{\frac{n+m-2}{1/n+1/m}} \frac{\overline{X} - \overline{Y}}{\sqrt{(n-1)S_X^2 + (m-1)S_Y^2}} .$$

What is the probability distribution of Z ?
2. Propose a test of $H_0 =$"$m_1 = m_2$" against $H_1 =$"$m_1 \neq m_2$".

Exercise 7.7.4 *Distance estimate in a Gaussian model.*

Let $Y_i = (Y_{1,i}, Y_{2,i})$, $i = 1, \ldots, n$ be a n-sample of Gaussian vectors with expectation $\mathbf{E}Y = (m_1, m_2)$ and with unknown covariance matrix. Let $d = \sqrt{m_1^2 + m_2^2}$ be the distance to the origin. Assume $d > 0$. Propose an estimate of d and give an asymptotical confidence interval.

Exercise 7.7.5 *Fisher information of a Bernoulli distribution.*

Let $X_1, \ldots, X_n$ be a n-sample of Bernoulli distribution of parameter $0 < \theta < 1$.

1. Compute the Fisher information.
2. Let $\overline{X}$ be the empirical expectation of $X_1, \ldots, X_n$. Is this estimate efficient?

Exercise 7.7.6 *Non-uniqueness of likelihood estimates.*

Let $X_1, \ldots, X_n$ be a n-sample of distribution density $p(x) = Ce^{-|x-\theta|}$.

1. Compute C.
2. Is the likelihood estimate of θ unique?

Exercise 7.7.7 *Integral approximation.*

Let $X_1, \ldots, X_n$ be an n-sample of a uniform distribution on $A \subset \mathbb{R}^k$. Let ϕ be a continuous function from A to $\mathbb{R}$.

1. Set $Y_i = \phi(X_i)$. Compute $\mathbf{E}Y_i$.
2. Let $S_n = \dfrac{1}{n} \sum\limits_{i=1}^{n} Y_i$. What is the limit of S_n as $n \to +\infty$? Give an asymptotical confidence interval on $\int_A \phi$ based on S_n as $n \to +\infty$.
3. Compare with the estimate given by a Riemann sum.

Exercise 7.7.8 *Uniform distribution.*

Let $X_1, \ldots, X_n$ be an n-sample of uniform distribution on $[0, \theta_0]$. θ_0 is unknown.

1. Compute the likelihood estimate $\widehat{\theta}_n$. Prove that this estimate is biased.
2. Set $\tilde{\theta}_n = \dfrac{n+1}{n}\widehat{\theta}_n$. Compute $\mathrm{var}(\tilde{\theta}_n)$. Is there a contradiction with the Cramér-Rao inequality?
3. Compute $\mathrm{var}(\overline{X})$. Compare with $\mathrm{var}(\tilde{\theta}_n)$.

Exercise 7.7.9 *Uniform distribution (ctd.).*

Let $X_1, \ldots, X_n$ be a n-sample of uniform distribution on $[\theta_1, \theta_2]$. $\theta_1 \geq 0$ is unknown and θ_2 is known. Propose and study a test $H_0 = "\theta_1 = 0"$ against $H_1 = "\theta_1 > 0"$.

Exercise 7.7.10 *Exponential models.*

Let f be a sigma-finite measure and T a random vector of $\mathbb{R}^k$ Consider the set $\Theta \subset \mathbb{R}^k$ of θ such that $exp < \theta, T >$ is f-integrable. Define:

$$\psi(\theta) = Log \int exp < \theta, T > df ,$$

for $\theta \in \Theta$.

1. Check that the measure $\mathbf{P}_\theta = exp(-\psi(\theta) + <\theta, T>)f$ is a probability distribution. Let $\mathbf{E}_\theta$ be the associated expectation.
2. Show that:

$$\mathrm{grad}\psi(\theta) = \mathbf{E}_\theta(T) ,$$
$$\frac{\partial^2 \psi}{\partial \theta_i \partial \theta_j} = cov_\theta(T_i, T_j) .$$

3. Compute the Fisher information matrix.
4. Compute the Kullback information between two distributions with parameters θ_1 and θ_2. Make a Taylor expansion of order 2 and compare with the Fisher information.
5. Show that the following distributions define exponential models: binomial distributions, Poisson distributions, exponential distributions, Gaussian distributions.

Exercise 7.7.11 *Sequence alignment (from [40], [41]).*

Consider a finite set $A = \{a_1, \ldots, a_r\}$. Classical biological examples are the nucleotides ($r = 4$), the amino acids ($r = 20$), and the codons ($r = 61$). Consider a probability distribution $\mathbf{p} = \{p_1, \ldots, p_r\}$. A random sequence S is sampled independently from A with probability $\mathbf{p}$. We associate with each a_i a score s_i. We require at least one score to be positive and the expected score $E = \sum_1^r p_i s_i$ to be negative.

1. Show that if the scores are of a likelihood-type (*i.e.* $s_i = \log q_i/p_i$, $q_i \geq 0$, $\sum_1^r q_i = 1$), then $E \leq 0$ holds automatically.

2. Show that there exists a unique $\lambda^\star > 0$ such that:

$$\sum_1^r p_i e^{\lambda^\star s_i} = 1 \, .$$

3. a) Let $X_i, i = 1, \ldots, n$ be n i.i.d. exponential random variables with parameter λ. Set $Y = \max_{i=1,\ldots,n} X_i$. Find $\lim_{n \to +\infty} \mathbf{P}\left(Y - \frac{\log n}{\lambda} \leq x\right)$.

   b) Let $M(n)$ be the segment of the sequence S with greatest additive score. This segment is usually called the maximal segment score. The study of the distribution of $M(n)$ is intricate (the previous question on the asymptotic distribution of Y is only a rough indication on the behavior of $M(n)$). One can prove (see [40, 41] for details and proofs) that there exists a constant $K^\star$ such that:

$$\lim_{n \to +\infty} \mathbf{P}\left(M(n) - \frac{\log n}{\lambda^\star} \leq x\right) = \exp\left(-K^\star e^{-\lambda^\star x}\right) \, .$$

Consider now two sequences S and S' of sizes n and n' sampled from A with probabilities $\mathbf{p}$ and $\mathbf{p}'$. Let M be the high maximal subalignment score between the two sequences S and S'. Let α ($0 < \alpha < 1$) be a given level. Let $x^\star$ such that:

$$\exp(-K^\star \exp(-\lambda^\star x^\star)) = \alpha \, .$$

Why can we say that any alignment of segments from two sequences has an high score (of level α) if M exceed $\frac{\log(nn')}{\lambda^\star} + x^\star$?

Exercise 7.7.12 *Herbivore-plankton model.*

1. Consider an aquatic herbivorous population eating phyto-plankton. Let $P(t)$ (resp. $H(t)$) the biomass at time t. The relationships are modeled by:

$$\frac{dP(t)}{dt} = rP(t)\left((K - P(t)) - \frac{BH(t)}{C + P(t)}\right) ,$$

$$\frac{dH(t)}{dt} = DH(t)\left(\frac{P(t)}{C + P(t)} - AH(t)\right) .$$

Show that there exists a steady state in the quadrant $x > 0$, $y > 0$ and that this steady state can be stable or unstable according to the parameters.

2. We want to estimate the parameters A and C. The experimenter can keep the biomass of phyto-plankton constant while some herbivores are introduced in the experimental pool. The system then evolves into its steady state. The biomass Q of herbivores is measured. The experiments are done in independent pools. In the i-th pool, a biomass m_i is kept and a biomass Q_i of herbivores is measured. Consider the model:

$$Q_i = \frac{m_i}{A(C + m_i)} + \varepsilon_i , \quad i = 1, \ldots, n ,$$

where the ε_i are centered Gaussian i.i.d. r.v. with unknown variance.
a) Justify this model.
b) Why do we need $n \geq 3$?
c) Write the likelihood equations.

Exercise 7.7.13 *Estimate for differential equations.*

Let f be a C^1 function. Consider the differential equation:

$$\frac{dx(t)}{dt} = f(x(t)) ,$$

with initial condition $x(0)$. Let $x^\star$ such that $f(x^\star) = 0$ and $f'(x^\star) < 0$. Let $x(t)$ be a solution such that:

$$\lim_{t \to +\infty} x(t) = x^\star .$$

The $y_n, n \geq 0$ are observed:

$$y_n = x(n) + \varepsilon_n .$$

The ε_n are centered second-order i.i.d. r.v. The following estimate of $x^\star$ is proposed:

$$\widehat{x_n} = \frac{1}{n}\sum_{k=1}^{n} y_n .$$

1. What is the (a.s.) limit of $\widehat{x_n}$ as $n \to +\infty$?
2. What is the limiting distribution of $\sqrt{n}(\widehat{x_n} - x^\star)$ as $n \to +\infty$?

Exercise 7.7.14 *Censored data.*

Consider a population that can either die after an accident, or a disease. If an individual dies after a disease, the age i when dying is modeled by the r.v. X_i. If an individual dies after an accident, the age i when dying is modeled by the r.v. C_i. To sum up, an individual dies at an age modeled by the r.v. $T_i = \inf(X_i, C_i)$. We assume that the individuals are independent and identically distributed.

1. Set $\delta_i = \mathbf{1}_{X_i \leq C_i}$. What does δ_i mean?
2. Assume that the variables X_i and C_i admit densities:

$$P(X_i \geq x) = \int_x^{+\infty} f_X(u)du \, ,$$

$$P(C_i \geq x) = \int_x^{+\infty} f_C(u)du \, .$$

Set

$$\lambda_X(x) = \frac{f_X(x)}{\int_x^{+\infty} f_X(u)du} \, ,$$

$$\lambda_C(x) = \frac{f_C(x)}{\int_x^{+\infty} f_C(u)du} \, .$$

What is the meaning of λ_X and λ_C?
3. From now on, we assume the function $\lambda_X(x)$ (resp. $\lambda_C(x)$) to be constant and equal to λ_X (resp. λ_C). What are the distributions of X_i, C_i and T_i?
4. We estimate λ_X by:

$$\widehat{\lambda}_X = \frac{\sum_{i=1}^n \delta_i}{\sum_{i=1}^n \delta_i T_i} \, .$$

Why? Is this estimate consistent?
5. Then we propose:

$$\tilde{\lambda}_X = \frac{\sum_{i=1}^n \delta_i}{\sum_{i=1}^n T_i} \, .$$

Is this estimate consistent? What is your own conclusion?

A

Appendices

A.1 Ordinary differential equations

Let I be an interval of the form $[t_0, T]$, $[t_0, T)$ or $[t_0, +\infty)$. Let f be a continuous function from $\mathbf{R}^m$ into $\mathbf{R}^m$ and let $y_0 \in \mathbf{R}^m$. We are looking for a continuous differentiable function y, defined from I into $\mathbf{R}^m$, such that, for every $t \in I$:

$$y'(t) = f(y(t)) , \qquad (A.1)$$
$$y(t_0) = y_0 .$$

The equation (A.1) is a first order differential equation. Let us recall that a p-order equation like

$$z^{(p)}(t) = \phi(z(t), z'(t), \dots, z^{(p-1)}(t)) ,$$

is amenable into a first order equation like (A.1) by the transformation $y_1(t) = z(t)$, $y_2(t) = z'(t)$, $\dots$, $y_p(t) = z^{(p-1)}(t)$.

Theorem A.1.1 *Cauchy-Lipschitz.*
If the function f satisfies the Lipschitz condition:

$$|f(y) - f(z)| \leq L|y - z| ,$$

for all $(y, z) \in \mathbf{R}^{2m}$, then the problem (A.1) has a unique solution.

Definition A.1.1 *Trajectory. A trajectory is the set $\{y(t), t \in I\}$, where y is a solution of (A.1).*

Definition A.1.2 *Stability.*
Let $I = [t_0, +\infty[$. A solution ψ of (A.1) is called a stable solution if, for all $\varepsilon > 0$, there exists $\delta > 0$ such that, for any solution ϕ of (A.1) satisfying:

$$|\phi(t_0) - \psi(t_0)| \leq \delta ,$$

we have, for all $t \geq t_0$,

$$|\phi(t) - \psi(t)| \leq \varepsilon \, .$$

Moreover if we have $\lim\limits_{t \to +\infty} |\phi(t) - \psi(t)| = 0$, *then we say that the solution ψ is asymptotically stable.*

We are interested in the steady states, defined by:

$$f(y^\star) = 0 \, .$$

The problem (A.1) with initial condition $y(t_0) = y^\star$ admits the solution $y(t) \equiv y^\star$. Therefore, we will speak, with a minor abuse of language, of the stability of $y^\star$.

The following results concern the stability of the solution near a steady state $y^\star$. Without loss of generality, we will assume that $y^\star = 0$.

A.1.1 Stability when $m = 1$

Assume that (A.1) takes the form:

$$y'(t) = ay(t) + g(y(t)) \, ,$$

where $g(y) = O(|y|^{1+\varepsilon}), \varepsilon > 0$, as $y \to 0$. Then:

1. $a > 0$. 0 is unstable.
2. $a < 0$. 0 is asymptotically stable.

A.1.2 Global behavior when $m = 1$

Let us consider the ordinary differential equation:

$$y'(t) = f(y(t)) \, ,$$

with initial condition $y(t_0) = y_0$. The qualitative study of the trajectory is done the following way.

1. Assume $f(y_0) < 0$. Denote, when it exists, $y^\star = \sup\{y \leq y_0, \ f(y) = 0\}$. The trajectory coming from y_0 cannot come back to y_0; this is forbidden by the Cauchy-Lipschitz Theorem that ensures the uniqueness of the solution. For the same reason, the trajectory is not allowed to cross the point $y^\star$. The trajectory remains into the interval $[y^\star, y_0]$. On this interval $[y^\star, y_0]$, the function f is negative. The derivative of the function y is therefore negative. Function y is decreasing and bounded. Function y converges to a limit as $t \to +\infty$. This limit has to be a steady point. If $y^\star$ does not exist, we show that the trajectory converges to $-\infty$ as $t \to +\infty$ with the same arguments.

2. Assume $f(y_0) > 0$. The same arguments prove that the trajectory converges, as $t \to +\infty$, to the smallest zero of function f that is greater than y_0, when it exists. Else, the trajectory converges as $t \to +\infty$ to $+\infty$.
3. Assume $f(y_0) = 0$. This is a steady point and the trajectory remains on this point.

Especially, an oscillating or asymptotically oscillating behavior is not possible in one dimension.

A.1.3 Stability when $m = 2$

Local behavior of a linear system.

Consider the linear system:

$$y_1' = ay_1 + by_2 , \tag{A.2}$$
$$y_2' = cy_1 + dy_2 .$$

Let A be the matrix $\begin{pmatrix} a & b \\ c & d \end{pmatrix}$. This matrix is usually called the stability matrix. Assume $\det(A) \neq 0$. Let λ, μ the eigenvalues of A. These eigenvalues can be real or complex numbers. If $\lambda = \alpha + i\beta$ (α, β real numbers, $\beta \neq 0$) is a complex number, then $\mu = \alpha - i\beta$ is the other eigenvalue. There exists a real non-singular matrix T such that $J = TAT^{-1}$ has one of the following forms:

$$J = \begin{pmatrix} \lambda & 0 \\ 0 & \lambda \end{pmatrix} \quad \lambda \neq 0 \tag{A.3}$$

$$J = \begin{pmatrix} \lambda & 0 \\ 0 & \mu \end{pmatrix} \quad \mu < \lambda < 0 \text{ or } 0 < \mu < \lambda \tag{A.4}$$

$$J = \begin{pmatrix} \lambda & 0 \\ \gamma & \lambda \end{pmatrix} \quad \lambda \neq 0, \ \gamma > 0 \tag{A.5}$$

$$J = \begin{pmatrix} \lambda & 0 \\ 0 & \mu \end{pmatrix} \quad \lambda < 0 < \mu \tag{A.6}$$

$$J = \begin{pmatrix} \alpha & \beta \\ -\beta & \alpha \end{pmatrix} \quad \alpha \neq 0, \ \beta \neq 0 \tag{A.7}$$

$$J = \begin{pmatrix} 0 & \beta \\ -\beta & 0 \end{pmatrix} \quad \beta \neq 0 . \tag{A.8}$$

The local behavior near $(0,0)$ and the usual terminology are given by the following figures.

1. Case A.3
   a) $\lambda < 0$.
   b) $\lambda > 0$.

Fig. A.1. Stable proper node

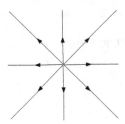

Fig. A.2. Unstable proper node

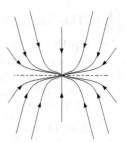

Fig. A.3. Stable node

2. Case A.4
   a) $\mu < \lambda < 0$.
   b) $0 < \mu < \lambda$.

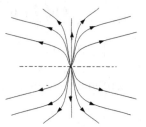

Fig. A.4. Unstable node

3. Case A.5
   a) $\lambda < 0$.

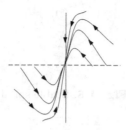

Fig. A.5. Stable improper node

b) $\lambda > 0$.

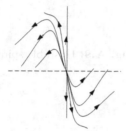

Fig. A.6. Unstable improper node

4. Case A.6.

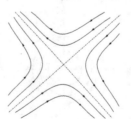

Fig. A.7. Saddle point

5. Case A.7
   a) $\alpha < 0$, $\beta < 0$.
   b) $\alpha > 0$, $\beta < 0$.
6. Case A.8
   a) $\beta < 0$.
   b) $\beta > 0$.

Fig. A.8. Stable spiral

Fig. A.9. Unstable spiral

Fig. A.10. Elliptic fixed point

Fig. A.11. Elliptic fixed point

Local behavior of a non-linear system

Consider the non-linear system:

$$y_1' = ay_1 + by_2 + g_1(y_1, y_2) \,,$$
$$y_2' = cy_1 + dy_2 + g_2(y_1, y_2) \,. \qquad (A.9)$$

Set $r = \sqrt{y_1^2 + y_2^2}$. Assume the existence of $\varepsilon > 0$ such that, near $(0,0)$, we have $g_1(y_1, y_2) = O(r^{1+\varepsilon})$, $g_2(y_1, y_2) = O(r^{1+\varepsilon})$ and such that $\dfrac{\partial g_1}{\partial y_2}$ and $\dfrac{\partial g_2}{\partial y_1}$ exist and are continuous on a neighborhood of $(0,0)$.

The local behavior of the non-linear system (A.9) can be deduced from the local behavior of the associated linear system (A.2).

1. A node for A.2 remains a node for A.9 and keeps the same stability.
2. A proper node for A.2 remains a proper node for A.9 and keeps the same stability.
3. An elliptic fixed point for A.2 becomes an elliptic fixed point or a node (stable or unstable) for A.9.
4. If A.2 is an unstable improper node, then every trajectory of A.9 converges (or keeps away from) to the origin with an angle of 0, $\pi/2$, π or $3\pi/2$ with the axis of x; the stability is kept.
5. If A.2 is a saddle point, then there exists a trajectory converging to the origin with an angle of 0, one converging to the origin with angle of π, the other keeping away from the origin.

A.1.4 Global behavior when $m = 2$

So far, we only considered the local behavior of the differential equations near the steady state. These behaviors have been given when $m = 2$, but can be easily generalized when $m > 2$. We will now give some global properties. We must keep in mind that these global properties are not available when $m > 2$ anymore.

Consider the following system:

$$y_1' = f_1(y_1, y_2),$$
$$y_2' = f_2(y_1, y_2).$$

Assume that $f = (f_1, f_2)$ is defined and continuous on an open bounded domain D of $\mathbf{R}^2$. Recall that a point $y^\star$ such that $f(y^\star) = 0$ is called a steady state and that a point such that $f(y^\star) \neq 0$ is called a regular point.

Definition A.1.3 *Limit point.*
  A point Q is a limit point of the trajectory C if there exists a sequence t_n, with $\lim\limits_{n \to +\infty} t_n = +\infty$, such that $(y_1(t_n), y_2(t_n))$ converges to Q as $n \to +\infty$. The set of limit points Q of C is denoted by $L(C)$.

Theorem A.1.2 *Poincaré-Bendixson.*
  Assume C to be contained in a closed subset $K \subset D$. If $L(C)$ only contains regular points, then

  1. either $C = L(C)$ and C is a periodic trajectory;
  2. either $L(C)$ is a periodic trajectory. We say then that C is a limit cycle.

Theorem A.1.3 *Classification of limit trajectories.*
Assume C being contained in a closed subset $K \subset D$. Assume that D only contains a finite number of steady states, then:

1. *either $L(C)$ is reduced to a unique steady point, and C converges to this steady point as $t \to +\infty$;*
2. *either $L(C)$ is a periodic trajectory;*
3. *either $L(C)$ contains a finite number of steady states and a set of trajectories, each of them converging to a steady state as $t \to +\infty$.*

Nota-bene : See [14] or [33] for a general theory on ordinary differential equations.

A.2 Evolution equations

The study of partial differential equations highly exceeds the outline of this book. The aim of this appendix is only to give some general ideas on evolution equation like the Fisher equation. For instance we refer to [10] for questions related to functional analysis, to [74] for numerical approximations, and [13], [28] and [44] for general results.

A.2.1 General problem

Our aim is to find a function $u(t, x)$ from $[0, T] \times \Omega$ into $\mathbb{R}$, such that:

$$\begin{cases} \frac{\partial u}{\partial t} = Au + f(t, x, u), & t \in [0, T], \quad x \in \Omega, \\ u(t, x) = 0, & t \in [0, T], \quad x \in \partial\Omega, \\ u(0, x) = \phi(x), \end{cases} \tag{A.10}$$

where Ω is an open bounded domain of $\mathbb{R}^N$ [1], with a "smooth" boundary[2], A an operator and ϕ a "smooth" function [3].

We will give the main outline when A is the Laplacian $A \equiv \Delta = \displaystyle\sum_{i=1}^{N} \frac{\partial^2}{\partial x_i^2}$.

Equation (A.10) is then called reaction-diffusion equation.

A.2.2 Homogeneous linear problem

Let us start with some classical results on finite dimensional system of differential equations:

---

[1] Assuming Ω bounded is not an actual limitation.
[2] for instance C^1 piecewise.
[3] For instance continuous and vanishing on $\partial\Omega$.

$$\begin{cases} \frac{d\overrightarrow{u}}{dt} = M\overrightarrow{u} \,, \\ \overrightarrow{u}(0) = \overrightarrow{u_0} \,, \end{cases} \tag{A.11}$$

where M is a matrix.

We can define the matrix exponential by writing the solution of (A.11) with the form $\overrightarrow{u(t)} = \exp(tM)\overrightarrow{u_0}$. Then the matrix exponential satisfies $\frac{d}{dt}\exp(tM) = M\exp(tM)$.

Consider equation (A.10) without the right side:

$$\begin{cases} \frac{\partial u}{\partial t} = Au \,, \quad t > 0, \quad x \in \Omega \,, \\ u(t,x) = 0, \quad t > 0, \quad x \in \partial\Omega \,, \\ u(0,x) = \phi(x) \,. \end{cases} \tag{A.12}$$

With suitable conditions on the operator A, satisfied by the Laplacian[4], we prove the existence and uniqueness [5] of (A.12). This solution is a global one with respect to t since the problem is a linear one.

Following the analogy with the matrix exponential, we can write the exponential operator as the solution of (A.12). The solution of (A.12) then becomes $u = \exp(tA)\phi$. The exponential operator satisfies $\frac{d}{dt}\exp(tA) = A\exp(tA)$.

In the case of the Laplacian, denote by $(e_i)_{i\geq 1}$ a Hilbert basis of $L^2(\Omega)$ where the e_i are eigenfunctions of $-\Delta$ that vanish on $\partial\Omega$:

$$\begin{cases} -\Delta e_i = \lambda_i e_i \text{ on } \Omega \,, \\ e_i \quad = 0 \text{ on } \partial\Omega \,. \end{cases}$$

The eigenvalues $(\lambda_i)_{i\geq 1}$ are positive. Denote by $<,>$ the inner product of $L^2(\Omega)$.

We are looking for solutions of the form:

$$u(t,x) = \sum_{i\geq 1} a_i(t)e_i(x) \,.$$

We can easily deduce $a_i(t) = a_i(0)\exp(-\lambda_i t)$. The constants $a_i(0)$ are fixed by the relation $\phi(x) = \sum_{i\geq 1} a_i(0)e_i(x)$, i.e. $a_i(0) = <\phi, e_i>$. The solution of (A.12) can therefore be written:

$$u(t,x) = \sum_{i\geq 1} a_i(0)\exp(-\lambda_i t)e_i(x) \,. \tag{A.13}$$

We can use the exponential operator, so (A.13) becomes:

$$u = \exp(t\Delta)\phi \,.$$

---

[4] One can for instance prove that $-A$ is a self-adjoint maximal monoton operator (e.g. [10, Ch.7]).

[5] Clearly, for a rigorous presentation, one needs to specify to which functional space the solution belongs.

A.2.3 Non-homogeneous linear problem

Consider equation (A.10) when the right side is independent of u:

$$\begin{cases} \frac{\partial u}{\partial t} = Au + f(t,x), & t \in [0,T], \quad x \in \Omega, \\ u(t,x) = 0, & t \in [0,T], \quad x \in \partial\Omega, \\ u(0,x) = \phi(x). \end{cases} \tag{A.14}$$

A method which is similar to the method of variation of parameters for ordinary differential equations, even technically more difficult, allows to exhibit a solution (that will be unique for the same reasons as previously). In the case of the Laplacian, as one can check it formally, this solution becomes:

$$u(t,x) = \sum_{i \geq 1} a_i(0) \exp(-\lambda_i t) e_i(x) \tag{A.15}$$

$$+ \sum_{i \geq 1} \int_0^t < f, e_i > \exp(-\lambda_i(t-s)) ds \, e_i(x).$$

Using $e^{t\Delta}$, (A.15) becomes:

$$u = \exp(t\Delta)\phi + \int_0^t \exp((t-s)\Delta) f(s,x) ds.$$

A.2.4 Back to the general problem

Function f is now depending on function u. From (A.15), we introduce the operator $\mathcal{T}$:

$$\mathcal{T}u = \sum_{i \geq 1} a_i(0) \exp(-\lambda_i t) e_i(x)$$

$$+ \sum_{i \geq 1} \int_0^t < f, e_i > \exp(-\lambda_i(t-s)) ds \, e_i(x). \tag{A.16}$$

Using notation $\exp(t\Delta)$, (A.16) becomes:

$$\mathcal{T}u = \exp(t\Delta)\phi + \int_0^t \exp((t-s)\Delta) f(s,x,u(s,x)) ds,$$

where function f is defined by (A.10) and depends on u. If f is locally Lipschitz, the operator $\mathcal{T}$, for T small enough, is a contraction. A fixed point Theorem in Banach spaces ensures the existence of an unique solution of $\mathcal{T}u = u$, that is of (A.10). Of course this solution is local with respect to time. A global control on f (e.g. [28, Ch.3]) transforms this local solution into a global one; a linear growth of f is sufficient.

A.2.5 Maximum principle

Let us come back to the linear homogeneous problem (A.12). If $\phi \in L^2(\Omega)$, we have the essential inequalities (*e.g.* [10, Ch. X.2]) for the solution $u(t, x)$ of (A.12):

$$\min(0, \inf_{\Omega} \phi) \leq u(t, x) \leq \max(0, \sup_{\Omega} \phi) .$$

Especially if the initial condition ϕ is positive and bounded, then the solution of (A.12) remains positive and does not explode.

Note that the maximum principle is valid for more general evolution equations (*e.g.* [28, Ch.3]).

A.3 Probability

A.3.1 Inequalities

Let X be a real random variable (in short r.v.) We have the following inequalities.

1. Markov inequality. .
   Let $\alpha > 0$ and $\varepsilon > 0$. If $\mathbf{E}|X|^\alpha < \infty$, then:

   $$\mathbf{P}(|X| \geq \varepsilon) \leq \frac{1}{\varepsilon^\alpha} \mathbf{E}|X|^\alpha .$$

2. Bienaymé-Tchebicheff inequality .
   Let $\varepsilon > 0$. If $\mathrm{var}(X) < \infty$, then:

   $$\mathbf{P}(|X - \mathbf{E}X| \geq \varepsilon) \leq \frac{1}{\varepsilon^2} \mathrm{var}(X) .$$

3. Jensen inequality .
   Let ϕ be a convex function from $\mathbb{R}$ into $\mathbb{R}$. If $\mathbf{E}X < \infty$ and $\mathbf{E}\phi(X) < \infty$, then:

   $$\phi(\mathbf{E}X) \leq \mathbf{E}\phi(X) .$$

4. Hölder inequality .
   Let p and q such that $p > 1$, $q > 1$ and $\dfrac{1}{p} + \dfrac{1}{q} = 1$.
   If $\mathbf{E}|X|^p < \infty$ and $\mathbf{E}|X|^q < \infty$, then:

   $$\mathbf{E}|XY| \leq \left(\mathbf{E}|X|^p\right)^{\frac{1}{p}} \left(\mathbf{E}|X|^q\right)^{\frac{1}{q}} .$$

   Cauchy-Schwarz inequality corresponds to the case $p = q = 2$.
5. Cramér inequality.
   Assume that the Laplace transform $t \to \mathbf{E}(\exp(tX))$ is defined on a neighborhood of 0. Then, for all $x > \mathbf{E}X$:

   $$\mathbf{P}(X \geq x) \leq \inf_t (\exp(-xt)\mathbf{E}(\exp(tX))) .$$

A.3.2 Convergences

Let X be a real r.v. and $X_n, n \geq 0$ a sequence of real r.v. The following type of convergences can be defined as $n \to +\infty$:

1. Almost sure convergence (in short a.s.).

$$X_n \to X \ (a.s.) \Leftrightarrow \mathbf{P}(\omega, \ X_n(\omega) \to X(\omega)) = 1$$
$$\Leftrightarrow \forall \varepsilon > 0, \ \mathbf{P}(\limsup |X_n - X| > \varepsilon) = 0 \ .$$

2. Convergence in probability.

$$X_n \to X \ (\mathbf{P}) \Leftrightarrow \forall \varepsilon > 0, \ \lim \mathbf{P}(|X_n - X| > \varepsilon) = 0 \ .$$

3. Convergence in distribution. We assume that $\mathbf{P}(X = x) = 0$ for all x.

$$X_n \to X \ (\mathcal{L}) \Leftrightarrow \forall x, \ \lim \mathbf{P}(X_n < x) = \mathbf{P}(X < x) \ .$$

4. Convergence L^p $(p \geq 1)$.

$$X_n \to X \ (L^p) \Leftrightarrow \lim \mathbf{E}|X_n - X|^p = 0 \ .$$

These convergences are related to each other the following way:

- A.s. convergence implies convergence in probability.
- Convergence L^p implies convergence L^q for $p \geq q \geq 1$.
- Convergence L^p, $p \geq 1$, implies convergence in probability.
- Convergence in probability implies convergence in distribution.

Lemma A.3.1 *Borel-Cantelli Lemma.*
Let A_n, $n \geq 0$ be a sequence of events.
If $\displaystyle\sum_{n \geq 0} \mathbf{P}(A_n) < \infty$ then $\mathbf{P}(\limsup A_n) = 0$.

A.3.3 Zero-one law

Let $X_1, X_2, \ldots, X_n, \ldots$ be a sequence of independent r.v. Let $\mathcal{A}_p$ be the sigma-algebra generated by $(X_n)_{n \geq p}$. Let $\mathcal{A}_\infty = \displaystyle\bigcap_{p \geq 1} \mathcal{A}_p$. Then, for all $F \in \mathcal{A}_\infty$,

$\mathbf{P}(F) = 0$ or $\mathbf{P}(F) = 1$.

A.3.4 Independent random variables

Let $X_1, X_2, \ldots, X_n, \ldots$ be a sequence of real independent and identically distributed (in short i.i.d.) r.v.

Theorem A.3.1 *Strong law of large numbers.*
If $\mathbf{E}|X_1| < \infty$, then, as $n \to +\infty$, $\displaystyle\frac{1}{n} \sum_{i=1}^{n} X_i$ converges (a.s.) to $\mathbf{E}X_1$.

Theorem A.3.2 *Central Limit Theorem.*

If $\mathbf{E}X_1^2 < \infty$, *then, as* $n \to +\infty$, $\sqrt{n}\left(\dfrac{1}{n}\displaystyle\sum_{i=1}^{n} X_i - \mathbf{E}X_1\right)$ *converges in distribution to a centered Gaussian r.v. with variance* $var(X_1)$.

Now assume that the Laplace transform of X_1 is defined on a neighborhoud of 0. Define the Cramér transform $h(x)$ of X_1:

$$h(x) = \inf_t(\log \mathbf{E}(\exp(tX_1)) - xt) .$$

Theorem A.3.3 *Large deviations (Chernoff).*
Let $x > \mathbf{E}X_1$. *Then, as* $n \to +\infty$:

$$\lim_n \frac{1}{n} \log \mathbf{P}\left(\sum_{i=1}^{n} X_i \geq nx\right) = h(x) .$$

A.3.5 Discrete-time martingales

Definition A.3.1 *Martingale.*
Let $X_1, X_2, \ldots$ *be a sequence of real r.v. defined on a probability space* $(\Omega, \mathcal{A}, \mathbf{P})$. *Let* $\mathcal{F}_1, \mathcal{F}_2, \ldots$ *be an increasing sequence of sigma-algebras of* $\mathcal{A}$. *The process* $(X_n)_{n\geq 1}$ *is called a martingale (with respect to the sequence* $\mathcal{F}_n$*) if the three following conditions are valid:*

1. $\mathbf{E}|X_n| < +\infty$ *for all* $n \geq 1$.
2. X_n *is* $\mathcal{F}_n$-*measurable for all* $n \geq 1$.
3. $\mathbf{E}(X_{n+1}|\mathcal{F}_n) = X_n$ *for all* $n \geq 1$.

Theorem A.3.4 *Convergence of martingales.*
If $\sup_{n\geq 1} \mathbf{E}X_n^2 < +\infty$, *then* X_n *converges in* L^2 *and a.s. to a squared integrable r.v. as* $n \to +\infty$.

Nota-bene : See [6], [9] and [42, 43] for general results in probability.

A.4 Statistics

A.4.1 Gaussian samples

Theorem A.4.1 *Gaussian samples.*
Let $X_1, \ldots, X_n$ *be a n-sample of Gaussian r.v. with expectation* m *and variance* σ^2. *Let* $\overline{X} = \dfrac{1}{n}\displaystyle\sum_{i=1}^{n} X_i$ *be the empirical expectation and*

$S^2 = \dfrac{1}{n-1}\displaystyle\sum_{i=1}^{n}(X_i - \overline{X})^2$ *be the empirical variance. Then* $\overline{X}$ *and* S^2 *are independent and:*

$$\sqrt{n}\frac{\overline{X}-m}{\sigma} \sim \mathcal{N}(0,1) ,$$

$$\frac{1}{\sigma^2}\sum_{i=1}^{n}(X_i - m)^2 \sim \chi^2(n) ,$$

$$\frac{1}{\sigma^2}\sum_{i=1}^{n}(X_i - \overline{X})^2 \sim \chi^2(n-1) ,$$

$$\sqrt{n}\frac{(\overline{X}-m)}{S} \sim t(n-1) ,$$

where $\chi^2(n)$ is the Chi-square distribution with n degrees of freedom and $t(n)$ is the Student distribution with n degrees of freedom.

A.4.2 Chi-square tests

Chi-square test for goodness of fit

Let X be a r.v. taking values in the set $\{1,\ldots,k\}$ and with probability distribution $p = p(1),\ldots,p(k)$. We want to test H_0 = "the distribution of X is $p^\star = p^\star(1),\ldots,p^\star(k)$" against H_1 "the distribution of X is different from $p^\star$". Let $X_1,\ldots,X_n$ be a n-sample of r.v. having the probability distribution $p = p(1),\ldots,p(k)$. Let $N_i = \sum_{j=1}^{n}\mathbf{1}_{X_j=i}$ and $\hat{p}_n(i) = \dfrac{N_i}{n}$. Set:

$$\chi_n^2(p^\star,\hat{p}_n) = n\sum_{i=1}^{k}\frac{(p^\star(i) - \hat{p}_n(i))^2}{p^\star(i)} .$$

Theorem A.4.2 *Chi-square test for goodness of fit.*
The sequence $\chi_n^2(p,\hat{p}_n)$ converges in distribution, as $n \to +\infty$, to a $\chi^2(k-1)$.

The acceptance region of H_0 with asymptotical level α is:

$$\chi_n^2(p^\star,\hat{p}_n) \leq \chi_\alpha^2(k-1) .$$

Chi-square test of independence

Let X and Y be two r.v. taking values in the sets $\{1,\ldots,k\}$ and $\{1,\ldots,l\}$. Let $p = \{p(i,j), 1 \leq i \leq k, 1 \leq j \leq l\}$ be the probability distribution of (X,Y). We want to test H_0 = "X and Y are independent" against H_1 = "X and Y are not independent". Let $(X_i,Y_i)_{1\leq i\leq n}$ be a n-sample of r.v. having the probability distribution $p = \{p(i,j), 1 \leq i \leq k, 1 \leq j \leq l\}$. Let $N_{i,j}$

be the number of observations of (i, j) in the sequence $(X_m, Y_m)_{1 \le m \le n}$. Let $\widehat{p}_n(i, j) = \dfrac{N_{i,j}}{n}$ and $\tilde{p}_n(i, j) = \dfrac{N_i N_j}{n^2}$. Set:

$$\chi_n^2(\tilde{p}_n, \widehat{p}_n) = n \sum_{i,j=1}^{k,l} \frac{(\widehat{p}_n(i, j) - \tilde{p}_n(i, j))^2}{\tilde{p}_n(i, j)} \, .$$

Theorem A.4.3 *Chi-square test of independence.*
The sequence $\chi_n^2(\tilde{p}_n, \widehat{p}_n)$ converges in distribution, as $n \to +\infty$, to a $\chi^2((k-1)(l-1))$.

The acceptance region of H_0 with asymptotical level α is:

$$\chi_n^2(\tilde{p}_n, \widehat{p}_n) \le \chi_\alpha^2(k-1, l-1) \, .$$

Chi-square test of symmetry

Let us take the same notations with $k = l$. Now we want to test H_0 = "the distributions $p(i, j)$ and $p(j, i)$ are identical" against H_1 = "the distributions $p(i, j)$ and $p(j, i)$ are different". Set:

$$\chi^2(\widehat{p}_n) = n \sum_{i,j=1}^{k} \frac{(\widehat{p}_n(i, j) - \widehat{p}_n(j, i))^2}{\widehat{p}_n(i, j)} \, .$$

Theorem A.4.4 *Chi-square test of symmetry.*
The sequence $\chi^2(\widehat{p}_n)$ converges in distribution, as $n \to +\infty$, to a $\chi^2\left(\dfrac{k(k-1)}{2}\right)$.

The acceptance region of H_0 with asymptotical level α is:

$$\chi_n^2(\widehat{p}_n) \le \chi_\alpha^2\left(\frac{k(k-1)}{2}\right) \, .$$

References

1. D.G. Aronson and H.F. Weinberger. Multidimensional nonlinear diffusions arising in population genetics. *Adv. Math.*, 30:33–76, 1978.
2. K.B. Athreya and P.E. Ney. *Branching Processes*. Springer-Verlag, 1972.
3. J.-P. Aubin. *Optima and Equilibria: An Introduction to Nonlinear Analysis*. Springer, 1998.
4. C. Auer. Dynamik von Laerchenwickler-Populationen laengs des Alpenbogens (in german). *Eidg. Anst. fuer Forstl. Versuchsweses*, 53:71–105, 1977. Fasc. 2.
5. J.D. Biggins. Chernoff's theorem in the branching random walk. *J. Appl. Prob.*, 14:630–636, 1977.
6. P. Billingsley. *Convergence of Probability Measures*. Wiley, New-York, 1968.
7. M. Bramson. *Convergence of solutions of the Kolmogorov equation to travelling waves*, volume 44. Memoirs Amer. Math. Soc., 1983.
8. M. Braun. *Differential Equations and Their Applications*. Springer-Verlag, 1983.
9. L. Breiman. *Probability*. Addison Wesley, 1968.
10. H. Brezis. *Analyse fonctionnelle, théorie et applications*. Masson, Paris, 1983.
11. S. Broadbent and J. Hammersley. Percolation processes I. Crystals and mazes. *Proc. Cam. Phil. Soc.*, 53:629–641, 1957.
12. H. Caswell. *Matrix Population Models*. Sinauer, 2001.
13. T. Cazenave and A. Haraux. *An Introduction to Semilinear Evolution Equations*. Oxford Lecture Series in Mathematics and Its Applications, Oxford University Press, 1998.
14. E.A. Coddington and N. Levinson. *Theory of Ordinary Differential Equations*. TMH Edition, 1972.
15. I. Cornfeld, S. Fomin, and I. Sinai. *Ergodic Theory*. Springer-Verlag, 1982.
16. R.F. Costantino, R.A. Desharnais, J.M. Cushing, and Brian Dennis. Chaotic dynamics in an insect population. *Science*, 275:389–391, 1997.
17. H. Cramer. *The Elements of Probability Theory*. Wiley, 1955.
18. D. Dacunha-Castelle and M. Duflo. *Probability and Statistics*, volume 1. Springer-Verlag, 1985.
19. D. Dacunha-Castelle and M. Duflo. *Probability and Statistics*, volume 2. Springer-Verlag, 1985.
20. J. Diamond. *Guns, Germs and Steel: the Fates of Human Societies*. Random House, 1997.
21. K. Falconer. *Fractal Geometry*. Lecture Notes, Monograph Series, Wiley, 1990.

22. M.J. Feigenbaum. Quantitative universality for a class of nonlinear transformations. *J. Stat. Phys.*, 19:25–52, 1978.
23. P.-H. Gouyon, J.-P. Henry, and J. Arnould. *Gene Avatars. The Neo-Darwinian Theory of Evolution.* Kluwer, 2002.
24. G. Grimmett. *Percolation.* Springer-Verlag, 1999. 2nd ed.
25. P. Guttorp. *Statistical Inference for Branching Processes.* Wiley Series in Probability and Mathematical Statistics, 1991.
26. J.B.S. Haldane. The combination of linkage values and the calculation of distances between the loci of linked factors. *J. of Genetics*, 8:299–309, 1919.
27. J. Hawkes. Trees generated by a simple branching process. *J. Lond. Math. Soc.*, 24:373–384, 1981.
28. D. Henry. *Geometric Theory of Semilinear Parabolic Equations.* Lecture Notes in Mathematics, Springer-Verlag, 1981.
29. P. Hoel, S. Port, and J. Stone. *Introduction to Stochastic Processes.* Houghton Mifflin, 1972.
30. J. Hofbauer and K. Sigmund. *Evolutionary Games and Population Dynamics.* Cambridge University Press, 1998.
31. S. Huet, A. Bouvier, M.A. Poursat, and E. Jolivet. *Statistical Tools for Nonlinear Regression.* Springer-Verlag, 2003. 2nd ed.
32. N. Ikeda and S. Watanabe. *Stochastic Differential Equations and Diffusion Processes.* North-Holland, 1989.
33. G. Iooss and D. Joseph. *Elementary Stability and Bifurcation Theory.* Undergraduate Texts in Mathematics, Springer-Verlag, 1990. 2nd ed.
34. K. Itô and H. McKean. *Diffusion Processes and Their Sample Paths.* Springer-Verlag, 1996. Reprint of 1st ed. (1974).
35. C. Jacob and J. Peccoud. Theoretical uncertainty of measurements using quantitative polymerase. *Biophys. J.*, 71:101–108, 1996.
36. P. Jagers. *Branching Processes with Biological Applications.* Wiley Series in Probability and Statistics, 1975.
37. E. Jolivet. *Introduction aux modèles mathématiques en biologie.* Masson, 1983.
38. B. Jordan. *Les imposteurs de la génétique.* Seuil, 2000.
39. T. Jukes and C. Cantor. Evolution of protein molecules. In *Mammalian Protein Metabolism*, pages 21–123. Academic Press, Ed. Munro,H., 1969.
40. S. Karlin and S. Altschul. Method for assessing the statistical significance of molecular sequence features by using general scoring schemes. *Proc. Natl. Acad. Sci. USA*, 87:2264–2268, 1990.
41. S. Karlin and A. Dembo. Limit distributions of maximal segmental score among Markov-dependent partial sums. *Adv. Appl. Prob.*, 24:113–140, 1992.
42. S. Karlin and H. Taylor. *A First Course in Stochastic Processes.* Academic Press, 1975.
43. S. Karlin and H. Taylor. *A Second Course in Stochastic Processes.* Academic Press, 1981.
44. O. Kavian. *Introduction à la théorie des points critiques.* Mathématiques & Applications, Springer-Verlag, 1993.
45. W.O. Kermack and A.G. McKendrick. Contributions to the mathematical theory of epidemics. *Proc. Roy. Soc. Ser. A*, 115:700–721, 1927.
46. W.O. Kermack and A.G. McKendrick. Contributions to the mathematical theory of epidemics. *Proc. Roy. Soc.*, 138:55–83, 1932.
47. W.O. Kermack and A.G. McKendrick. Contributions to the mathematical theory of epidemics. *Proc. Roy. Soc.*, 141:94–122, 1933.

48. P. Kloeden and E. Platen. *Numerical Solution of Stochastic Differential Equations*. Springer-Verlag, 1992.
49. A.N. Kolgomorov, I.G. Petrovsky, and N.S. Piskunov. Etude de l'équation de la diffusion avec croissance de la quantité de matière et son application à un problème biologique. *Bjul. Moskowskogo Gos. Univ., Série internationale A 1*, pages 1–26, 1937.
50. M. Kot, M. Lewis, and P. Van Den Driessche. Dispersal data and the spread of invading organisms. *Ecology*, 77:2027–2047, 1996.
51. E.S. Lander and D. Botstein. Mapping Mendelian factors underlying quantitative traits using RFLP linkage maps. *Genetics*, 121:185–199, 1989.
52. C. Laredo and J. Pernes. Models for pearl millet domestication as an example of cereal domestication. *J. Theor. Biology*, 131:289–305, 1988.
53. C. Laredo and A. Rouault. Grandes déviations, dynamique de population et phénomèmes malthusiens. *Ann. Inst. Poincaré.*, 19:323–350, 1983.
54. V. Le Corre. *Organisation de la diversité génétique et histoire post-glaciaire des chênes blancs européens: approche expérimentale et par simulation*. PhD thesis, Institut national agronomique, 1997.
55. T.Y. Li and J.A. Yorke. Period three implies chaos. *Amer. Math. Monthly*, 82:985–992, 1975.
56. A. Lotka. *Elements of Physical Biology*. Williams and Wilkins, Baltimore, 1925. (Reprinted in 1956), Elements of Mathematical Biology. Dover Publications (New York)).
57. D. Ludwig, D.G. Aronson, and H.F. Weinberger. Spatial patterning of the spruce budworm. *J. Math. Biology*, 8:217–258, 1979.
58. D. Ludwig, D.D. Jones, and C.S. Holling. Qualitative analysis of insect outbreak systems: the spruce budworm and forest. *J. Anim. Ecol.*, 47:315–332, 1978.
59. R. May. Biological populations with nonoverlapping generations: Stable points, stables cycles, and chaos. *Science*, 186:645–647, 1974.
60. R. May and G. Oster. Bifurcations and dynamic complexity in simple ecological models. *Am. Nat.*, 110:573–599, 1976.
61. J. Maynard-Smith. *Evolution and the Theory of Games*. Cambridge University Press, 1982.
62. S. Meyn and R. Tweedie. *Markov Chains and Stochastic Stability*. Springer-Verlag, 1993.
63. K. Mullis, F. Ferre, and A. Gibbs. *The Polymerase Chain Reaction*. Birkhäuser, Cambridge, MA, 1994.
64. J.D. Murray. *Mathematical Biology*. Springer-Verlag, 1990.
65. J. Nedelman, P. Heagerty, and C. Lawrence. Quantitative PCR: procedure and precision. *Bull. Math. Biol.*, 54:477–502, 1992.
66. J. Nedelman, P. Heagerty, and C. Lawrence. Quantitative PCR with internal control. *Comput. Appl. Biosci.*, 8:65–70, 1992.
67. E. Odum. *Fundamentals of Ecology*. Saunders, 1971.
68. B. Øksendal. *Stochastic Differential Equations*. Springer-Verlag, 6th ed., (1st ed., 1985), 2003.
69. A. Okubo. *Diffusion and Ecological Problems: Mathematical Models*, volume 10. Springer-Verlag, 1980.
70. A. Okubo and S. Levin. *Diffusion and Ecological Problems: Modern Perspectives*. Springer-Verlag, 2001.
71. G. Owen. *Game Theory*. Academic Press, 2nd edition, 1982.

72. J. Pernes. La génétique de la domestication des céréales. *La Recherche*, 146:910–919, 1983.

73. B. Prum, F. Rodolphe, and E. de Turckheim. Finding words with unexpected frequencies in deoxyribonucleic acid sequences. *J. R. Statist. Soc. B*, 57:205–220, 1995.

74. P.A. Raviart and J.M. Thomas. *Introduction à l'analyse des équations aux dérivées partielles*. Masson, 1983.

75. T. Royama. *Analytical Population Dynamics*. Chapman & Hall, 1992.

76. G. Ruget. Large deviations and more or less rare events in population dynamics. In *Rhythms in Biology and Other Fields of Application*, pages 388–400. Springer-Verlag, Proceedings Luminy 1981, Eds M.Cosnard, J.Demongeot and A. Le Breton, 1981.

77. A.N. Sarkovsky. Coexistence of cycles of a continuous map of a line into itself (in russian). *Ukr. Mat. Z.*, 16:61–71, 1964.

78. D.H. Sattinger. Stability of waves of nonlinear parabolic systems. *Adv. Math.*, 22:312–355, 1976.

79. R. Schinazi. *Classical and Spatial Stochastic Processes*. Birkhäuser, 1999.

80. R. Schonmann. On the behavior of some cellular automata related to bootstrap percolation. *Ann. Proba.*, 20:174–193, 1992.

81. S. Sorin. *A First Course on Zero-Sum Repeated Games*, volume 37. Mathématiques & Applications, Springer-Verlag, 2002.

82. M. Tomasini. Etude de la vitesse de colonisation par une population en expansion. Rapport de stage, INRA, 1997.

83. P. Verhulst. Notice sur la loi que la population suit dans son accroissement. *Corr. Math. et Phys.*, 10:113–121, 1838.

84. V. Volterra. Variazioni fluttuazioni del numero d'individui in specie animali conviventi. *Mem. Acad. Lincei*, 2:31–113, 1926. (Variations and fluctuations of a number of individuals in animal species living together. Translation in R. N. Chapman. 1931. Animal Ecology. McGraw-Hill, New York).

85. J. Von Neumann and O. Morgenstern. *Theory of Games and Economic Behaviour*. Princeton University Press, Princeton, 1944.

86. J. Watson, N. Hopkins, J. Roberts, J. Steitz, and A. Weiner. *Molecular Biology of the Gene*. Benjamin/Cummings, 1987.

87. J. Weibull. *Evolutionary Game Theory*. MIT Press, 1995.

List of Figures

Index

Universitext

Aguilar, M.; Gitler, S.; Prieto, C.: Algebraic Topology from a Homotopical Viewpoint

Aksoy, A.; Khamsi, M. A.: Methods in Fixed Point Theory

Alevras, D.; Padberg M. W.: Linear Optimization and Extensions

Andersson, M.: Topics in Complex Analysis

Aoki, M.: State Space Modeling of Time Series

Arnold, V. I.: Lectures on Partial Differential Equations

Audin, M.: Geometry

Aupetit, B.: A Primer on Spectral Theory

Bachem, A.; Kern, W.: Linear Programming Duality

Bachmann, G.; Narici, L.; Beckenstein, E.: Fourier and Wavelet Analysis

Badescu, L.: Algebraic Surfaces

Balakrishnan, R.; Ranganathan, K.: A Textbook of Graph Theory

Balser, W.: Formal Power Series and Linear Systems of Meromorphic Ordinary Differential Equations

Bapat, R.B.: Linear Algebra and Linear Models

Benedetti, R.; Petronio, C.: Lectures on Hyperbolic Geometry

Benth, F. E.: Option Theory with Stochastic Analysis

Berberian, S. K.: Fundamentals of Real Analysis

Berger, M.: Geometry I, and II

Bliedtner, J.; Hansen, W.: Potential Theory

Blowey, J. F.; Coleman, J. P.; Craig, A. W. (Eds.): Theory and Numerics of Differential Equations

Blyth, T. S.: Lattices and Ordered Algebraic Structures

Börger, E.; Grädel, E.; Gurevich, Y.: The Classical Decision Problem

Böttcher, A; Silbermann, B.: Introduction to Large Truncated Toeplitz Matrices

Boltyanski, V.; Martini, H.; Soltan, P. S.: Excursions into Combinatorial Geometry

Boltyanskii, V. G.; Efremovich, V. A.: Intuitive Combinatorial Topology

Bonnans, J. F.; Gilbert, J. C.; Lemaréchal, C.; Sagastizábal, C. A.: Numerical Optimization

Booss, B.; Bleecker, D. D.: Topology and Analysis

Borkar, V. S.: Probability Theory

Brunt B. van: The Calculus of Variations

Carleson, L.; Gamelin, T. W.: Complex Dynamics

Cecil, T. E.: Lie Sphere Geometry: With Applications of Submanifolds

Chae, S. B.: Lebesgue Integration

Chandrasekharan, K.: Classical Fourier Transform

Charlap, L. S.: Bieberbach Groups and Flat Manifolds

Chern, S.: Complex Manifolds without Potential Theory

Chorin, A. J.; Marsden, J. E.: Mathematical Introduction to Fluid Mechanics

Cohn, H.: A Classical Invitation to Algebraic Numbers and Class Fields

Curtis, M. L.: Abstract Linear Algebra

Curtis, M. L.: Matrix Groups

Cyganowski, S.; Kloeden, P.; Ombach, J.: From Elementary Probability to Stochastic Differential Equations with MAPLE

Dalen, D. van: Logic and Structure

Das, A.: The Special Theory of Relativity: A Mathematical Exposition

Debarre, O.: Higher-Dimensional Algebraic Geometry

Deitmar, A.: A First Course in Harmonic Analysis

Demazure, M.: Bifurcations and Catastrophes